RIDING
THE
HIGH WIRE

T0308741

RIDING
THE
HIGH WIRE

AERIAL MINE TRAMWAYS IN THE WEST

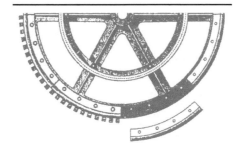

ROBERT A. TRENNERT

UNIVERSITY PRESS OF COLORADO

Copyright © 2001 by the University Press of Colorado

Published by the University Press of Colorado
5589 Arapahoe Avenue, Suite 206C
Boulder, Colorado 80303

The University Press of Colorado is a cooperative publishing enterprise supported, in part, by Adams State College, Colorado State University, Fort Lewis College, Mesa State College, Metropolitan State College of Denver, University of Colorado, University of Northern Colorado, University of Southern Colorado, and Western State College of Colorado.

The paper used in this publication meets the minimum requirements of the American National Standard for Information Sciences—Permanence of Paper for Printed Library Materials. ANSI Z39.48-1992

Library of Congress Cataloging-in-Publication Data

Trennert, Robert A.
  Riding the high wire : aerial mine tramways in the West / Robert A. Trennert.
    p. cm.
Includes bibliographical references and index.
  ISBN 0-87081-630-6 (alk. paper) — ISBN 0-87081-631-4 (pbk. : alk. paper)
  1. Mine haulage—West (U.S.)—History—19th century. 2. Mine haulage—West (U.S.)—History—20th century. 3. Aerial tramways—West (U.S.)—History—19th century. 4. Aerial tramways—West (U.S.)—History—20th century. I. Title.

  TN332 .T74 2001
  622'.66—dc21

                                                          2001002382

Cover design by Laura Furney
Text design by Daniel Pratt

CONTENTS

PREFACE

This project grew out of a long fascination with mining history. Since my days in high school, I have visited and photographed ghost towns and mining camps. Much of my interest also focused on railroads and their relationship with mining. In the course of these journeys I took note of the ghostly remains of several aerial tramway systems, although I really did not appreciate their importance to mining history. In 1997 I became president of the Mining History Association, and in searching around for a topic to discuss at the annual conference, I began to explore the history of aerial tramways. As I investigated the topic, I found that no overall history of these devices had ever been published despite the fact that during their heyday (1890–1920), tramways played a significant role in the operation of western mines.

As the history of a piece of technology, a number of limitations need to be observed. While considerable engineering data on tramways exists, it is not of much interest to the general reader and will not be detailed—this is available in several books.

Also, tramways were built for many purposes other than mining and existed elsewhere around the world. Although an interesting subject in their own right, tramways located outside the western United States and Canada or used for purposes such as logging are not discussed in detail. The focus, then, is on tramways constructed for the purpose of mining in the western portions of the Northern Hemisphere. Moreover, because literally hundreds of these devices existed in the West, no attempt is made to provide comprehensive coverage of every system.

Once I launched the project, I discovered that many people were interested in the subject and were willing to share their knowledge with me. I owe a deep debt of gratitude to many individuals, institutions, and corporations. Among the manufacturing companies that helped with the research are Interstate Equipment Corporation (especially Vice President Leo J. Vogel Jr.); Williamsport Wirerope Works, Inc.; and USX Corporation (formerly U.S. Steel). The staffs of many libraries and archives also provided a great deal of useful material. Of particular note are: Alaska State Library; Arizona Department of Mines and Mineral Resources; the Arizona Collection at Arizona State University Library; Bancroft Library; California Historical Society; Colorado Historical Society; Denver Public Library; Eastern California Museum; University of Idaho Library; Kootenay Museum Association; Michigan Technological University Archives and Copper Country Historical Collections; Mohave County Historical Society; San Juan Historical Society; Thomas Fisher Rare Book Library at the University of Toronto; Utah State Historical Society; and Wyoming State Archives. Thanks also go to Gary Krahenbuhl, Dean of the College of Liberal Arts and Sciences, and Noel Stowe, Chair of the Department of History, at Arizona State University for their support.

Quite a number of individuals went out of their way to provide personal help or to share information. Special thanks must go to: Bob Spude and Karl Gurcke of the National Park Service; Douglas R. Thayer, Murray Lundberg, Noel Kirschenbaum, David F. Myrick, Peter T. Hodge, Roger Burt, H. Mason Coggin, Ed Hunter, Stan and Sheldon Schwedler, Glen Crandall, Lynn R. Bailey, and Erik Nordberg. I am also deeply indebted to my graduate assistant, Kathleen L. Howard, and to my good friend and fellow historian, Duane A. Smith, for all their help.

RIDING
THE
HIGH WIRE

Introduction

A visitor to the mining camps of the Far West at the beginning of the twentieth century would in all probability have seen one or more aerial tramways in operation. Running day and night these devices formed an integral part of the mining industry, hauling ores from mines to reduction or transportation facilities and carrying supplies on the return trip. Many tramways were spectacular. They spanned gorges, rivers, and mountain ravines on strands of wire smaller than a broom handle. In an era before airplanes, persons riding the trams often expressed awe at being lifted hundreds of feet above the earth. One passenger pretty well summed up the experience in 1898: "Straight up the mountainside and into a dark canyon I went as if I were a bird. Higher and higher up from the ground the cables carried me, and I was afraid to look down."[1]

Mine operators, of course, expressed little concern for the scenic wonders of their machinery. They used ropeways because they provided an economical alternative to building roads or railroads by moving ores and supplies in a straight line over natural obstacles. As mining engineer T. A. Rickard, perhaps the most famous mining writer of the day, remarked in 1903, "These numerous aerial ropes spanning the intermountain spaces like great spiders' webs, are an important feature of mining in the San Juan [Colorado] region." Although tramways were considerably less glamorous than railroads, without them it would have been impossible to operate many of the most notable western mines. Such apparatuses, noted a 1908 issue of *The Mining and Metallurgical Journal*, "could not but appeal to progressive operators." They could soar over the most difficult terrain, did not require costly roadbeds or bridging, were unaffected by heavy snow and rain, and economically hauled large amounts of materials to and from mines.[2] The erection of an aerial tramway also offered visible evidence of prosperity and served as an inducement to speculate in company stock.

Aerial mine transportation did not just suddenly appear in western mining camps. The technology evolved over many decades and did not become available until after the western mineral rushes were well underway. As a consequence, the tramways were developed and improved upon very much in conjunction with the maturation of the frontier mining industry between 1870 and 1920. These devices represent just one of the hundreds of innovations of the era that rapidly transformed western mining from its reliance on the hand labor of individual men to the use of large-scale powered machinery. In this regard, the discussion of tramway development must be kept within the framework of other momentous mining advances. Aerial tramways, along with such things as hard-rock mining, pneumatic drills, the widespread application of electricity, and eventually methods of mass production—with such twentieth-century improvements as low-grade copper extraction, gold dredging, and gold heap leaching—all played a role in revolutionizing the industry. It should be noted, moreover, that these developments were part of a worldwide phenomenon that owed as much to European ingenuity as to American entrepreneurship.

The idea of moving materials by means of a rope stretched between two points in order to overcome natural obstacles dates back at least to the Middle Ages. During the seventeenth and eighteenth centuries, small ropeways are known to have existed in Europe, South America, and Africa. One crude system for transporting earth and sand was constructed in England during the early 1700s. It consisted of two parallel ropes running around hand-cranked pulleys at each end. By means of wicker baskets attached to the ropes, materials could be loaded and unloaded with the assistance of two men. Whether this device proved successful or not is unknown, although it was supposed to be more efficient than "the common way" of moving materials over distances as great as 500 yards.[3] Such early experiments were not suited for mining, primarily because the hemp ropes and supporting structures were incapable of handling heavy loads or covering significant distances.

The development of a practical way to manufacture wire rope set the stage for the construction of more useful aerial tramways. Small metal wires woven together in the form of a rope are known to have existed in ancient Egypt and there is some indication that a wire rope tramway was constructed in Germany as early as 1644. Nevertheless, it was not until the early part of

2

the nineteenth century that practical methods of manufacturing cable came about. The initial use of wire rope actually focused on suspension bridges and ship rigging. Credit for introducing wire rope to England is given to George Binks, who in 1830 proposed using wire cable in place of hemp rope on ships because it was stronger and lighter. About 1835, after convincing the Royal Navy to try his idea, Binks and George Harris established a small manufacturing plant at Great Grimsby. They were soon joined by Scottish inventor Andrew Smith, who secured several patents for "an improved mode of manufacturing bands, belts and straps to be employed in place of ropes and chains." These early wire ropes were handmade by setting spools of wire on a platform, then drawing it through a plate containing a series of holes. A revolving machine, which moved away from the plate on a track or sled, wove the strands into a rope much in the way hemp ropes were manufactured. This tedious process required the labor of several men, while producing only a limited amount of rope. Yet by 1840 wire rope factories of this sort existed in England, Sweden, and Germany.[4]

Three decades later several dozen factories were manufacturing wire rope in England. These operations turned out about 17,000 tons of wire rope annually, "in a multitude of forms, from one to many strands, in gauges as fine as a hair to as thick as a finger." In addition to cables patterned after hemp ropes, manufacturing companies also produced flat wire ropes. These were constructed from bunches of fine wires laid parallel and bound by wires that crossed the others at an angle, thereby forming the flattened rope. Various types of cables were thus available for a number of uses, including ship's rigging, telegraph lines, suspension bridges, towing hawsers, railway signals, and lightning conductors. They were also used for a variety of mining purposes, particularly as a replacement for hemp hoisting ropes.[5]

Meanwhile, inventors in America were developing their own wire rope products. John A. Roebling, a German immigrant who later became famous for building the Brooklyn Bridge (1883), is generally given the honor of first using this material. In 1832 Roebling helped found the town of Saxonberg, Pennsylvania. Soon thereafter he became interested in the operation of the nearby Portage Railway Incline Planes, which hauled goods over the Allegheny Mountains separating sections of the Pennsylvania Canal System. These inclines lifted cars of freight up steep grades by means of large hemp ropes, which regularly failed. According to Roebling's biographer the disastrous results of such failures sparked the "invention" of wire rope: "It occurred to Roebling that if a rope could be made of iron wire so as to be flexible enough to be wound on a windless, it should cost little more than a hemp cable but would possess much greater tensile strength with about one-fourth the diameter; and above all, it should outlast a dozen ropes woven from vegetable fiber."[6]

After some experimentation, in 1840 Roebling opened a rope factory on his farm at Saxonberg. Using a field almost a half-mile long, the young inventor, employing a hand-operated twisting machine, wove together lengths of wire to form a seven-strand rope, a process that he patented two years later. Roebling was eventually able to convince the portage railway to try his product, and soon thereafter he began to provide wire rope to American manufacturers of mining equipment, bridge builders, and shipwrights. As demand for wire rope increased, the budding industrialist realized that his primitive factory at Saxonberg had become obsolete.[7]

In 1848 Roebling accepted an invitation to move his operation to Trenton, New Jersey, partly through the efforts of Peter Cooper. Cooper, a well-known inventor, had been involved with many notable projects, including the construction of America's first railway locomotive, the "Tom Thumb," in 1830. By the 1840s he had become the nation's foremost manufacturer of iron, which led him to erect a rolling mill at Trenton in 1845 for the manufacture of rails and other products. Two years later Cooper, his son Edward, and son-in-law Abram S. Hewitt incorporated the Trenton Iron Company (which years later would become a leading producer of aerial tramways) to operate the mill. Another partnership, Cooper, Hewitt & Co., managed the company, then the largest of its type in the United States. Cooper may have expected Roebling to purchase wire products from his plant, but the ambitious German immigrant erected a mill of his own and began to draw wire. John A. Roebling's company quickly blossomed into a gigantic enterprise, manufacturing iron and steel wire and engaging in some of the most spectacular bridge projects of the nineteenth century.[8]

Roebling was well positioned to develop the first American aerial tramways, yet he never entered the business of fabricating these devices, although his wire rope products became an industry stalwart. In fact, it seems from the limited evidence available that his friend and competitor, Peter Cooper, built the first American aerial ropeways. In response to an 1872 *Engineering and Mining Journal* article claiming that "wire rope tramways" were a European invention, Cooper, Hewitt & Co. responded that Peter Cooper had constructed such a device in 1832 to transport landfill materials near Baltimore. If true, as the company claimed, then "the invention is of American origin, and . . . Peter Cooper is entitled to the credit of it." Unfortunately, there is little to support this claim. However, it is fair to conclude that Peter Cooper early became interested in the use of aerial cables to transport freight and materials.[9]

Cooper did construct several tramways during the 1850s. By this time the ironmaster operated several blast furnaces in New Jersey and Pennsylvania. To carry iron ore to these furnaces, he designed and constructed a two-mile tramway at Ringwood, New Jersey, in 1853. About the same time he erected a similar device to transport "iron ore, coal, and limestone to the top of a blast furnace near Phillipsburg, N.J." Being able to lift ore from canal barges directly to the furnaces by means of a bucket line enabled Cooper to produce some 25,000 tons of pig iron a year.[10] At that time, however, he seemed content to let these trams operate without applying his aerial technology to other uses.

Despite the pioneering efforts of Peter Cooper and others, the development of aerial tramways in Europe outpaced what was happening in the United States. Prior to 1870 several varieties of aerial transport were being used in the coal mines of France and Belgium. One report observed that at Prymont in the province of Savoy, "the owner of the asphalte quarries there has long employed wire-ropes for bringing down his bags of raw materials from the hill-side above to his works on the Rhone and from Modane the anthracite from the Alpine formations is brought down thus by a wire from the mouth of an adit, at a prodigious elevation above that town." Meanwhile, in 1856 Henry Robinson of Settle, England, secured a patent for a single-cable aerial ropeway to be used for the transportation of coal.[11] Although these European systems are noteworthy, they remained crude and experimental, with little in common aside from the use of wire rope to transport various materials.

Back in the United States, little further use of tramways is noted until after the Civil War, although they were surely used in ironworks, slate quarries, and possibly in eastern coal mines. The post-1865 boom in western mining, however, quickly created a need for economically feasible transportation systems to tap the remote but potentially rich mineral districts being developed by American enterprise. Little wonder, then, that western miners began to pay much more attention to the early European and American experiments. If perfected and standardized, aerial tramways offered many advantages.[12]

CHAPTER ONE

ANDREW S. HALLIDIE
AND THE ENDLESS WIRE ROPEWAY

The discovery of gold at Sutter's Mill by James Marshall in January 1848 changed American mining. It sparked the California gold rush, which in turn led to the development of fabulous mineral properties across the Far West. From the Great Basin and the Rocky Mountains to the Klondike and Alaska, prospectors and miners sought out the vast fortune of mineral wealth locked up by Mother Nature for millions of years. Indeed, from 1850 until the early decades of the twentieth century, mining opened the West. It sparked the growth of such major cities as San Francisco and Denver, encouraged the invention and manufacture of thousands of new technological devices, led to financial investment on an unprecedented scale, and changed the western landscape and the people living on it.

During the first decade after Marshall's discovery, perhaps three hundred thousand people came to California. Gold fever quickly spread into the interior as prospectors discovered mineral deposits all along the western slope of the Sierra Nevada range. Other entrepreneurs just as quickly went into the business of supplying miners with all their needs. The forty-niners benefited from good luck. They were able to recover large amounts of gold with little effort and almost no knowledge of geology or mining techniques because California's first gold rested in streambeds, the result of erosion that had washed it down from the mother lode. As a consequence, mining required

only a minimal outlay of money. All a man needed was time, a pan to separate the gold from the gravel, and perhaps a rocker, long tom (trough), or
sluice, all easily constructed. Of course, the supply of placer gold was quickly
depleted and the day of the individual prospector soon faded. He would be
replaced by eastern capitalists willing to finance hydraulic and, eventually,
quartz or hard-rock mining. Within a decade, mining made the transition
from small-scale surface efforts to a large industry, requiring expensive new
technology in order to extract even greater wealth from under the ground.[1]

Prospectors, displaced by these changes, soon fanned out across the West,
seeking new bonanzas and setting off additional rushes. New strikes came
slowly at first, but in 1859 miners from California discovered the silver-rich
Comstock Lode in what is now western Nevada. Meanwhile, the discovery of
gold deposits in Colorado (1859) drew nearly a hundred thousand fifty-niners
from the Mississippi Valley. Of particular interest to this study is the fact that
the Comstock and Colorado discoveries required lode mining. Outside capital soon came in to develop underground workings. As this occurred new
inventions and machines appeared on the market. In addition to the process
of extracting gold by amalgamation, "square set" timbering, pumps, hoists,
and more powerful explosives all came into general use. As Rodman W. Paul
has noted with regard to Nevada: "The Comstock's technical demands, then,
were met by a combination of innovation and adaptation from many parts of
the world. . . . A willingness to spend lavishly, and to try and then discard
expensive equipment, characterized Comstock operations."[2]

Although the early Comstock miners were faced with the expensive propo-
sition of having to transport ore by wagon from Virginia City to the mills
along the Carson River, no real alternative existed until after the end of the
Civil War. With the return of peacetime conditions, however, companies
financed with funds from San Francisco or New York entered western mining
in a big way, often building impressive mine properties in an effort to attract
more investors. In the case of Virginia City, topographical conditions and
the large number of mines made it possible to solve transportation problems
by constructing the Virginia & Truckee Railroad, completed in 1870.[3] Other
locations would not be as fortunate and would have to find cheaper answers
to the same problem.

This became particularly evident as interest in silver mining began to
spread from the Comstock into the remote Great Basin area of eastern Nevada.
At locations such as Oreana and Eureka experiments in smelting eventually
began to pay off. Along with better technology came new corporations with
enough resources to consolidate and develop prospective mine properties.[4]
As the output of precious metals (especially silver) increased, towns grew in
size and sophistication and the need for more efficient machinery became
obvious. In this respect, one of the most pressing needs in this rugged area
with no rail connections was for an effective technology to transport ores

from mines to mills, which, because of the need for water, were usually located some distance away.

During the 1860s San Francisco emerged as the manufacturing and supply center for the western mining industry. As Lynn R. Bailey has noted:
"California's isolation, coupled with complex mineralogical problems, which
in many cases had to be solved instantly, decreed that there would develop
on the West Coast an industry which would make San Francisco a center for
advancement of mining technology." In the shops and foundries of that city,
inventive westerners experimented with old European technologies as well as
new concepts, modifying everything to fit western conditions. A great deal of
innovative mining machinery soon began to appear. The list was almost endless: stamp mills, crushers, hoists, boilers, pumps, furnaces, compressors, and
smelter products.[5] In such an atmosphere, fueled by great demand and the
prospect of substantial profits, came the development of the first practical
aerial tramways adaptable to mining on the Pacific Slope.

Andrew Smith Hallidie, future developer of the San Francisco cable car
system, earned the distinction of developing the first successful western tramways, although it is clear that he built on the work of other inventors. Hallidie,
born in London in 1836, came by his interest in cable technology naturally.
His father, Andrew Smith, was one of the pioneers of wire rope manufacture
in England, being engaged in that business between 1835 and 1849. Coming
to California with his father during the gold rush, young Andrew first pur-
sued the life of a miner, embarking on a number of unproductive placer ventures before turning to various construction efforts. Recognizing the utility
of wire rope, Hallidie produced the first wire cable manufactured on the
Pacific Coast in 1856, and a year later the young entrepreneur opened a cable
factory, A. S. Hallidie & Company, in San Francisco. To make his product,
Hallidie reportedly melted down all the horseshoes he could buy from local
livery stables.[6]

The first wire produced by Hallidie was used to make cables for mine
hoists. These proved to be much stronger than hemp, yet flexible enough to
be wound around a winch. Most of his early cables were "flat iron" rope
about four inches wide by a half-inch thick. Being less likely to wear out and
having greater tensile strength, these cables became quite popular on the
Comstock. In 1869, for example, the Ophir Mine at Virginia City received
two such ropes, 1,500 feet long, that quickly proved their worth. During the
same period, Hallidie was also using wire rope to construct suspension bridges
along the Pacific Coast, a number of which were over 300 feet in length. As a
result of these achievements, Andrew Hallidie quickly emerged as one of San
Francisco's most successful industrialists. In 1868 he was honored by being
elected president of the prestigious Mechanic's Institute, a position he held
until 1877.[7] Although he could have successfully confined his business to the
production of wire rope and building bridges, as did his contemporary John

Andrew S. Hallidie (1836–1900). San Francisco civil leader and developer of the first successful wire rope tramway to be used at western mines. He is also noted for building San Francisco's famous cable car system. Courtesy California Historical Society, FN-25322.

Roebling, Hallidie's fascination with aerial transportation led him to experiment with cable tramways.

Hallidie probably began to test tramway theories in the late 1860s. He may have been influenced by reports coming from England about new applications. One article that reached the West Coast told of the construction of a cableway at the Leicestershire stone quarries. Some three miles in length, it carried cut stone from its source to a railway siding. Powered by steam, the line moved at a speed of six miles per hour and was said to be a complete success. English engineers were so impressed with the tramway (undoubtedly the brainchild of Charles Hodgson) that they began "discussing the possibility of constructing a tramway of this kind between Dover and Calais, which should be supported from a line of pillars in the channel, and along which passengers could be carried."[8] Although clearly a fanciful and impractical dream, Hallidie realized the value of more realistic applications.

The inventor's early ideas focused on what he called the "Endless Wire Ropeway," a single-line device that would receive a rope (spliced together to form a continuous loop) and then return it in the opposite direction. To support the cable as it traveled between large pulleys at each end, towers, usually no more than 250 feet apart, would be placed at appropriate locations across the intermediate terrain. A smooth run of the cable could be assured by attaching grooved pulleys or sheaves to the top of the towers. Thus, much like a modern ski lift, the cable moved in an endless circle. Bolted to the cable were small arms, which were connected to carrier buckets or hooks, to which supplies or ore bags might be secured.[9]

Before his invention could prove successful, Hallidie had to perfect a number of operating components. One involved devising a way for the large pulleys at each end to grip the rope without slipping so that the transmission of power could be regulated. This he overcame with the invention of an "improved grip pulley," which he patented on February 22, 1870. The new pulley featured a groove with jaws that compressed under strain, gripping the rope and preventing it from slipping, "but as soon as the strain is removed the jaws will work freely in their sockets, and allow the rope to open them, and consequently free itself from the pulley." Hallidie displayed his grip pulley at the 1869 Mechanic's Institute exhibit. Shortly thereafter he tackled another problem. As he explained, on earlier wire ropeways "it had been necessary to detach the cars or buckets from the rope before passing the end pulleys." To correct this, he developed a permanent arm (or clip) that could be attached to the cable in such a way that it easily passed through the pulleys. He also patented a dump car that could be tripped automatically and a method of permitting cars to change the angle of ascent or decent without problems.[10]

With these enhancements, Hallidie put his invention on the market. An article in the April 18, 1871, issue of San Francisco's *Mining and Scientific Press*

This artist's rendering of Hallidie's early single-rope tramway design appeared in the Engineering and Mining Journal, *July 2, 1872. Note that the ore carriers were suspended on a swivel between two carrier posts. It is uncertain if this carrier system ever entered service. Courtesy Michigan Technological University Archives and Copper Country Historical Collections, negative 05600.*

(no doubt supplied by Hallidie) detailed the entire system. Noting that the invention "is one of very considerable merit [and] of the greatest importance to miners," the story described the technology at length. Of note was the fact that tramways could be powered by gravity if the loads were descending, that carriers could be set up to automatically dump loads, and that costs were relatively low. Simple calculations showed an initial outlay of $5,000 per mile for the construction of a gravity-powered system, plus $7,000 per year in operating expenses, including labor. With such an outlay, a two-mile tramway, running at four miles per hour and using buckets capable of carrying fifty pounds of ore, could deliver ten and a half tons per hour.

The *Press* article also listed the advantages claimed by Hallidie:

> No grading or road building is required. It can work under all circumstances of weather, with great depths of snow on the ground, during heavy storms and freshets. It can run constantly without rest; as well during a dark night as a clear day. It can run across deep gorges and chasms. It can pass around precipitous bluffs and perpendicular cliffs. The ropes can never leave the posts or sheaves. It can furnish and transmit power when there is sufficient descent by its own generation, or by an engine attached

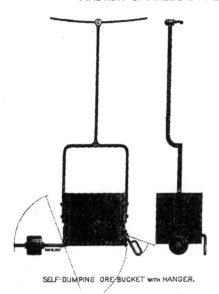

SELF-DUMPING ORE-BUCKET with HANGER.

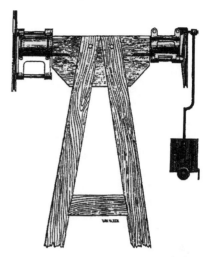

STATION-FRAME — SIDE-ELEVATION.

These drawings show three of the major components of Hallidie's tramway, including the distinctive grip pulley or "bull wheel," as they appeared in an 1878 catalog. They were used during the late nineteenth century on his single-rope systems. Courtesy of the Bancroft Library, University of California, Berkeley.

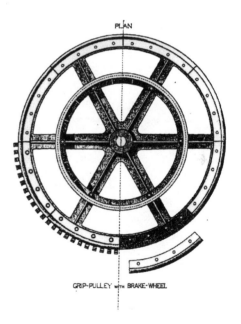

PLAN

GRIP-PULLEY with BRAKE-WHEEL

at either end. . . . By using the duplex carrier it can convey any material, such as lumber, goods, ores and even passengers, from place to place.[11]

Advertisements for Hallidie's Endless Wire Ropeway appeared in San Francisco as early as March 1871. Interested parties were instructed to contact David R. Smith, a civil and mechanical engineer employed by A. S.

Hallidie & Company, who would be charged with the actual construction while Hallidie worked on other inventions, including the beginnings of the San Francisco cable car system and additional tramway improvements.[12] Although mine operators soon began to express interest in his device, Hallidie would not erect the first aerial mine tramway in the West. That honor fell to Charles Hodgson of England.

Hodgson, an inventor from the town of Richmond in Surrey, had gained knowledge experimenting with single-cable ropeways during the 1860s. On July 20, 1868, he secured patent rights for an invention he called an "Improved Means of an Apparatus for Transporting Loads," much of which directly copied Henry Robinson's 1856 design. The following year Hodgson formed the Wire Tramway Company Ltd., in association with engineer William Thomas Henney Carrington. Miles Beale served as secretary. This company constructed two experimental tramways in 1869, the one near Leicester previously described and another at a Richmond gravel pit. In 1870 the company erected a five-mile experimental line at Brighton to study the possibility of opening a sixty-mile line (in five-mile segments) on the island of Ceylon.[13]

Hodgson's ropeway, sometimes called the English system, was a device similar to Hallidie's, although with some significant variations. His system also featured a single endless wire rope, but the carrier, or box, was attached by a different and much less secure method. The box-head (or saddle) was made of a notched hardwood block, lined with leather. It thus depended on friction to grip the rope. In order to pass over the towers the box-head had four small wheels "which run on rails placed on each support, and thus lift the box-head at these points from the rope and enable it to pass the pulleys, the momentum acquired keeping the load in motion until the box-head again catches the rope beyond the pulley." Wooden boxes capable of holding about 200 pounds were attached to the box-head by means of a curved arm. The system apparently had no automatic dumping features and could not operate over steep grades.[14]

Hodgson was looking for a way to give his tramway a practical test when developments in the White Pine Mining District of eastern Nevada came to his attention. White Pine, in the desert about thirty miles southeast of Eureka, had been the site of the rich Treasure Hill silver strike in early1868. In the rush that followed, miners and prospectors from across the West invaded the surrounding area. Nothing stopped this dazzling stampede, not even freezing weather or the lack of accommodations and supplies. Camped on the barren hills, suffering from disease, and feuding with each other, this desolate group sought their fortune. Within months thousands of claims were filed; cities such as Hamilton and Eberhardt came to life with scores of saloons, hotels, and brothels; over a hundred mines opened, and mills and smelters went up. During the height of production in 1869 and 1870, about $4 million worth of silver came out of the ground. Soon, however, produc-

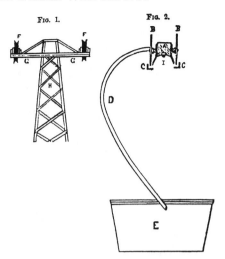

FIG. 1.   FIG. 2.

Drawings of the Hodgson tramway built at Treasure Hill, Nevada. Fig. 1 shows a standard tower. The pulleys (F) support the wire rope. In Fig. 2, the box-head (A) rests on the wire rope (I), as shown, and is carried along the moving rope. The box (E) is suspended from the box-head by a curved arm (D). The box-head has four wheels (B) that run on rails (C) as it crosses the tower. Mining and Scientific Press, *September 10, 1870.*

tion began to fall off and many of the less fortunate souls left for greener pastures. Although this ultimately proved to be the beginning of the end at White Pine, efforts to consolidate and streamline operations continued apace as optimism in some quarters remained quite high.[15]

One of the thorny problems at White Pine involved transporting ore from the mines on Treasure Hill to the mills below. Not only were a third of production costs consumed by hauling ores down the mountain by wagon, but the uncertain weather often stopped all movement. Teamsters were in such great demand that they could, and did, ignore the more inaccessible mines, leaving them to stockpile ore in hopes of finding transportation. In the midst of this situation, White Pine residents learned that a group of English investors had purchased five of the most promising mines, forming the Eberhardt and Aurora Mining Company Ltd. in early 1870. The company's directors hoped to turn a profit through consolidation of facilities, bringing in English miners to work under contract, and maximizing the production of ore. Additionally, because severe storms in early 1870 had stopped delivery of ore to the mill, the company decided to deal with this uncertainty by connecting their new International Mill to the mines with an aerial ropeway.[16]

Convinced that a tramway would be a good investment, Melville Atwood, a mining engineer employed by the company, contacted Hallidie in San Francisco. Hallidie provided Atwood with a pamphlet describing his tramway and estimated that a tramway at Treasure Hill, "with everything complete and in running order," would cost about $4,000 per mile. Impressed, but noncommittal, Atwood concluded that "I think it will answer our purpose for the present but not as a permanent work." The Eberhardt and Aurora's managers, however, preferred to use British technology. As a consequence, they entered into a contract with Hodgson's agent, William Thairwell & Company of Middlesex. Accompanied with drawings and instructions provided by the inventor, Thairwell journeyed to White Pine to oversee construction. Meanwhile, most of the wire and machinery arrived by ship from England through the port of San Francisco. When informed about the tramway in December 1870, local residents welcomed the news. Said the local newspaper, "The Eberhardt Co.'s Wire Tramway, which will pass the dumps of the principal mines on the southern slopes of the hill, will enable those mines also to deliver ore at the mills even when the roads are blocked up with snow." By the end of the month fifty men were employed in constructing the towers, stringing wire (which came in great coils), and installing the machinery. Predictions that the tramway would be in operation by early January seemed valid.[17]

The Treasure Hill tramway, 11,000 feet in length, represented a remarkable feat of engineering. It descended some 2,000 feet from the mines to the lower terminal at Eberhardt City and was powered by a sixteen-horsepower engine. The traveling wire was strung between sturdy wooden towers ranging in height from 20 to 130 feet. Construction of the $135,000 system did not go smoothly, however. Delays of an unknown character plagued the project, causing miners to complain that "again, ore is piling up awaiting the completion of the tramway." Not until mid-March was the first wire installed, but then stormy weather intervened. Finally, in early May engineers made an attempt to start the tramway, only to see their efforts foiled. After some hasty modifications, the tramway was finally placed in operation on or about May 20, 1871. As soon as test runs were completed, the tramway was put to work "transporting from the Ward Beecher mine to the International Mill about 125 tons of ore per day. . . . At the tramway station of the Beecher 30 men are employed, on the night and day shifts, in breaking down the ore from the immense dumps, and sacking and shipping it."[18]

The tramway experienced problems from the beginning. Much to the company's consternation, the one-inch cable was found to expand and contract because of temperature variations between the lower and upper terminals. This resulted in slack wire, causing the ore carriers to scrape the ground at several locations. The operation thus had to be halted until engineers could return to the site and make adjustments, which consisted of raising a few towers, more accurately measuring the length of the rope, and repositioning the tightening pulley from the upper terminal to the midway point, where the slack was the most severe. This adjustment seemed to work, and by September 1871 the tramway was restored to operation. Local accounts noted that the modified system carried 200 pounds of ore per bucket, at a cost ranging from seventy-five cents to one dollar per ton. Admitting that "it is a difficult matter to get it in successful operation," it was nevertheless proclaimed to be the longest tramway in the United States, answering "the purpose of a railway, without the necessity of embankments, tunnels, or bridges."[19]

Other technical problems proved more difficult to overcome. Given that the line ran over some "very rough, precipitous terrain," the wooden saddle grips either slipped or permitted the ore carriers to swing wildly, causing them to crash into the towers and dump valuable ore on the ground. This condition, which could not be solved, quickly caused the company to limit the tramway to low-grade ore and haul the more valuable product to the mill by wagon. In addition, the cable wore out within six months, forcing a costly delay while a new line was installed. Unfortunately, wrote engineer C. J. Bulkley, the new rope proved to be hard and brittle "and not so well adapted to the variable temperatures of this locality, as the old one." How much longer the tramway continued to operate is unclear. Like many boom situations, the Treasure Hill mines soon began to play out, although the Eberhardt and Aurora's British investors stuck it out until the end of the decade. But with declining profits and high operating costs, it seems unlikely that the company was willing to cover the expense of maintaining an inefficient tramway after 1873. Nevertheless, the Treasure Hill tramway demonstrated the potential of aerial transportation, even if the English design did not quite measure up to American conditions.[20]

While Hodgson experimented with the Treasure Hill tramway, Andrew Hallidie continued to improve his own system. That he was keenly aware of his competition could be seen in two patents secured on May 30, 1871, which corrected flaws inherent in his rival's design. In particular, Hallidie found a way to improve the wooden saddles so that they would firmly hold the rope when passing over towers. Although Hallidie's own tramway did not employ such saddles, his patent may have been intended to keep Hodgson from improving his system in the United States without paying a fee. Meanwhile, the San Francisco inventor continued to tinker with his own design, redesigning and realigning the weight-bearing pulleys to prevent a loaded carrier from pulling down the rope to such an extent that it would be unable to properly enter a tower pulley.[21]

It appears from the limited evidence available that Hallidie's first operational tramway was constructed in the Freiberg Mining District of eastern Nevada in 1872. The Freiberg silver deposits, discovered in 1856, were so isolated that little development took place prior to 1870. The reasons were obvious. Situated about eighty miles due south of the White Pine district in a dry and barren area on the northern slopes of the Worthington Mountains, the region lay miles from any kind of reliable transportation.[22] Nevertheless, the district caught the eye of Judge C. C. Goodwin, who in late 1871 purchased the Morning Star Mine, "said to be as rich as any in the district." Willing to gamble with investors' money, Goodwin constructed a mill in February 1872. In order to make this expensive operation more cost-efficient, he also contracted with Andrew Hallidie to install a 2,500 foot tramway at his mine. The young entrepreneur quickly fabricated the necessary machinery and wire rope, sent it to Freiberg, then headed there himself to supervise construction. Everything apparently went well and by the end of June the tramway entered service. As installed, the system was capable of discharging ten tons of ore per hour using the labor of just two men. As Goodwin wrote, "after several weeks trial upon our mine, the unanimous verdict of all who have seen it, is a complete unquestioned success." There also seemed to be every reason to believe that the operation could be extended and doubled in capacity if necessary.[23]

Hallidie used Judge Goodwin's endorsement to sell his system. He also distributed an artists' depiction of his tramway in operation, emphasizing its simplicity. Mining journals quickly endorsed the product. The *Mining and Scientific Press* observed that "there is little doubt but that Mr. Hallidie will be rewarded for his invention, for it is an apparatus which can be introduced with a great savings of time, labor, and money at many of the mines on the Coast. Its success, and the approbation of the owner of this mine, will do much towards introducing the tramway into general use." Although no comparisons were made with the Hodgson system, the mining journals clearly gave preferential coverage to the American system.[24]

With at least two aerial mine tramways in operation by 1872, what to call these inventions stirred debate in professional circles. Several different devices seemed to be using similar terminology. An example of this confusion involved the Stevens Mine in Colorado, which in 1869 had constructed what it referred to as a "Wire tram-way." In fact, the Stevens "tram-way" turned out to be an 867-foot incline railway that utilized cables to move ore cars down a steep grade. To clarify matters, the *Engineering and Mining Journal* suggested that the term "tramway" was inappropriate because that word already applied to railways. It thus seemed "better to put the word *air* in the name and call the apparatus, an air way, or something of that kind."[25] Nevertheless, the *Journal's* suggestion failed to catch on. Until the end of the century, when the name "aerial tramway" became more general, such terms as "cableway," "wire-rope-way," and "wire tramway" continued to confuse the situation.

Meanwhile, Hallidie turned his attention to the booming silver mines located in Little Cottonwood Canyon, Utah, some twenty-five miles southeast of Salt Lake City. Here he found conditions that seemed to beg for the construction of one or more of his endless wire ropeways. At the upper end of the narrow canyon lay Alta City, at an elevation of 8,200 feet. Situated around the town, especially on Emma Hill, were hundreds of mine sites, most notably the Emma Mine, which appeared to be fabulously wealthy. Because of the rough terrain, ore from the developed mines had to be hauled down a twisting mountain road at a cost of almost seven dollars per ton. Even more frustrating were the weather conditions: "The snow lies at Alta for about seven months and has a depth of from 8 to 20 feet deep for a distance of five miles below Alta. It is a region of avalanches and snow slides which, as is well known have several times occasioned serious damage and loss of life. The heavy grade in the cañon is unfavorable to the construction of a narrow-gauge railroad, and the depth of the snow necessitates snow sheds, etc., attended with great expense." All these conditions presented mine owners with considerable aggravation. Ore could only be moved during the summer months and the high cost of haulage precluded the ability to ship low-grade ores.[26]

During the summer of 1872 Hallidie entered into a contract to build a 2,400-foot tramway for the Vallejo Tunnel and Mine Company, an operation located next to the Emma Mine. Built in August 1872, using a wire rope five-eighths of an inch in diameter, the tramway descended some 600 feet and was capable of moving 100 tons of ore during a six-hour period. This tramway served several of the local operators, including the Emma Hill Consolidated Mining Company and the Flagstaff Mine. Mine managers seemed quite happy with the system. Two years after the line became operational, Emma Hill Consolidated superintendent L. U. Colbath wrote Hallidie that the apparatus "continues to work splendidly, and [with] but little wear on the rope. It has been everything that was promised, and has proved to be the cheapest way to move ores on steep mountain sides." Owners of the Vallejo Tunnel remarked that the system was strong and safe, with the rope suspended some forty feet above the ground, enabling the tramway to function all winter "without the slightest trouble."[27]

Although a small local system, the success of Hallidie's tramway kindled the interest of a group of Salt Lake City investors in constructing an eight-mile tramway to run all the way from Alta City to Granite at the mouth of the canyon where several new smelters were located. With the Emma Mine, now owned by British investors, still appearing to be "one of the great mines of the world," there seemed little doubt that a tramway would rake in the profits. Construction estimates, based on Hallidie's patent and including sixteen and a half miles of hardened and tempered steel wire, came in at $150,775. As the promoters noted in May 1873, "the undertaking if carried out will be of great benefit to the mining district of Little Cottonwood and will be [a] means of developing the mines more systematically and continuously than has been done before."[28]

Unfortunately, such optimism quicky turned sour when the news came in mid-1873 that the famous Emma Mine was in fact a giant swindle perpetuated on its British investors. Typical of speculative ventures in the West, company directors had promised fantastic dividends while glossing over skeptical reports in order to attract gullible men and women looking for quick profits. When the ore suddenly gave out, the bubble burst. News of the fraud caused mining activity in Little Cottonwood Canyon to collapse, along with the tramway project. Some of the mines were able to struggle along on a sporadic basis and the Vallejo's short tramway operated at least until 1874. Nonetheless, no one was interested in investing heavily in new projects. As a Salt Lake City newspaper remarked in January 1874, "the best of our mines are lying idle, either awaiting the arbitration of quarrels in England or the adjustment of lawsuits at home."[29]

For the next decade Andrew Hallidie held a virtual monopoly on the construction of western mine tramways. Although specific information is sparse, it appears that some two dozen such devices were installed by 1883. Known sites include the Chicago Mine in Utah, the Harley Mine at Kernville and the Standard Gold Mine at Bodie, California, the Blue Jacket Mine near Elko, Nevada, and some eight tramways in Colorado. Hallidie was able to dominate the business primarily because he controlled the production of wire rope on the Pacific Coast, especially after 1870 when he acquired the wire rope factory of Joshua Gray, his only competitor. Shortly thereafter, a number of prominent San Francisco businessmen, with Hallidie as their agent, organized the Pacific Wire Rope Manufacturing Company. By 1875 the factories of A. S. Hallidie and Pacific Wire combined employed fifty men and were producing some seventy-five tons of manufactured wire per month, most of which was used in the mining and maritime industries. On April 20, 1882, Hallidie combined his various interests into a single unit, the California Wire Works, capitalized at $500,000, with himself as the major stockholder. This company sold all types of wire rope and wire products, including wire tramways. Five months later the new company acquired a square block of land in North Beach and began the construction of a new factory, which, when completed in 1884, included a wire mill, wire rope factory, barbed-wire factory, machine shops, and warehouses.[30]

During these business moves, Hallidie continued to improve and promote his tramways. At the 1874 Mechanic's Fair in San Francisco he built a display model to demonstrate some of his recent improvements. In particular, he was eager to show off a new bucket design that emptied ore at the mill dumps and then automatically closed. Four years later, he prepared a twenty-five-page booklet describing the Endless Wire Ropeway and providing the reader with general instructions for its erection. This pamphlet, complete with drawings of the patented components and an artist's depiction of a mountain tramway, boasted that Hallidie's ropeway had been in operation for six years, proving "itself in every way the most reliable, economical, and simple mode of conveying ores." A number of testimonials from satisfied operators rounded out the promotional piece.[31]

Despite his early domination, challenges to Hallidie's tramway business began to appear in the mid-1880s. Aside from the introduction of the double-rope design that came later in the decade (see Chapter 2), other entrepreneurs attempted to cash in on the opportunities presented by an explosion of west-

FIG. 1. GENERAL VIEW OF ROPEWAY.

1878 artist's rendering of a Hallidie tramway, showing the design that actually went into service. Courtesy Bancroft Library, University of California, Berkeley.

ern mining activity prompted by the construction of railroads and more readily available financing. By 1884 John A. Roebling's Sons Company, the nation's leading producer of wire rope, had established a sales office in San Francisco in direct competition with the California Wire Works. Although Roebling's did not actually construct tramways, their product made it possible for others to avoid Hallidie's monopoly. The first notice of full-fledged competition entering the field came at the 1883 Denver National Exhibition of Mining and Industry, where inventor Charles M. Huson of St. Louis unveiled a single-rope tramway that featured automatic loading and unloading devices. Promoters claimed it to "be far superior to the great English tramway erected in White Pine, Nev., for the Eberhardt mine. This is so simple and safe that it will answer well for various transportation in places of dangerous travel, and across streams and chasms." Huson, in the field supervising construction of a tramway at the Sampson Mine near Gladstone, Colorado, apparently did not appear in person at the exhibition.[32]

Not to be outdone, Hallidie erected a demonstration tramway at the same exhibition, supervised by L. C. Trent, the Denver representative of Fraser & Chalmers, who served as Hallidie's agent in the Rockies. Trent boasted that Hallidie had already built eight tramways in Colorado, which were "in full and successful operation." Moreover, he claimed, Huson was a pirate, having infringed on Hallidie's patents in order to make his machine work. To back his claims for superiority, Trent supplied the press with a testimonial from John H. Gear, president of the Iowa and Consolidated Mining Company of Summitville, Colorado, which had recently installed a 3,700-foot tramway. "The tramway," Gear wrote, "works to a charm in every particular. It will transport to our mill in ten hours more than our mill, forty stamps, can crush in twenty-four hours. I would say, at a speed of 250 feet per minute, that is, if the buckets are fully loaded, it will carry nearly 200 tons in twenty-four hours. It beats bull teams badly."[33]

Despite claims of distinctiveness, both designs were relatively similar. The Huson system, however, did possess some advantages. These became evident in 1887 when the *Mining and Scientific Press* described a Huson tramway constructed at the Pay Rock Mine near Silver Plume, Colorado. Accompanied by a drawing illustrating the Pay Rock operation, the article described the loading and unloading features, which were a marked improvement over Huson's original 1882 patent. Fully automatic loading, never a strong suit of the Hallidie trams, was accomplished by having the empty bucket attach itself to a loading ore hopper as it rounded the terminal pulley. While still in motion, the ore was dumped into the bucket, then the hopper was cut away to be refilled before the next bucket arrived. To regulate the speed of the tramway during loading the upper terminal was provided with a hand brake. The operation thus required two men, one to load the hopper and one to regulate the line's speed. At the lower terminal, Huson installed a premanufactured box, built as a unit and pitched at an angle conforming to the slope of the mountain, that received the loaded buckets and by the action of a cam dumped the buckets as they rounded the end pulley. Other distinctive features of this system included grip wheels that, unlike Hallidie's patent, were adjustable to fit any size rope (needed since ropes were continually shrinking in diameter due to wear); a vertical clip that enabled "the rope to run in an open-grooved sheave of any depth"; and a distinctive style of tower in which the main support beams crossed each other.[34]

Although Huson focused most of his efforts in Colorado, he did open an office on Fremont Street in San Francisco, using the firm of Parke and Lacy to market tramways virtually in Hallidie's backyard. Nonetheless, Huson made his reputation in the Rockies. By the early 1890s his "Patent Automatic Wire Rope Tramway" was being manufactured and marketed by C. W. Badgley & Company of Denver. Their 1893 catalog, said to be in great demand, proclaimed "that we have erected, in the mining districts of the United States, more miles of tramway in the last three years than all other manufacturers of wire rope tramways combined." Huson based his success on simplicity, a major feature of which limited bucket capacity to no more than 150 pounds. According to the inventor, smaller buckets distributed weight more evenly over the entire line, thereby causing less wear and tear, and thus permitting the

This photo shows one of the early Huson tramways. Although not identified, comparison with a Mining and Scientific Press *drawing (January 1, 1887) indicates that it is the tramway constructed for the Pay Rock Mine at Silver Plume, Colorado. Note the substantial tower near the waterfall as well as the lighter ones carrying cable over the hill. Courtesy Center of Southwest Studies, Fort Lewis College, Durango, Colorado.*

Drawings of the Huson main loading (top) and unloading (bottom) mechanisms used by the company in its standard single-rope tramway. These devices were touted for their automatic loading and unloading as well as simplicity and ease of installation. Mining and Scientific Press, *January 1, 1887.*

system to be operated with greater speed and evenness. Another selling feature was making the installation so basic that any good mechanic could construct and repair a tramway: "no parts are so heavy or cumbersome that they cannot be packed to the most inaccessible points."[35]

The construction of single-rope tramways reached a peak during the 1890s. Huson dominated the Rocky Mountain area, building over a dozen lines in Colorado alone early in the decade. Many of his projects were connected in one way or another with the building of narrow-gauge railroads, which made mining in this rugged region more practical. Among his most notable efforts was the 2,279-foot tramway at the Sunnyside Extension Mine located at an altitude of 13,000 feet near Animas Forks in San Juan County. Although the arrival of the Denver & Rio Grande Railroad at Silverton in 1882, and the subsequent development of the Sunnyside Lode, unlocked great riches, ores still had to be packed to Silverton by mule train over difficult mountain trails. To ease one portion of this journey, mine superintendent Rasmus

Loading operation on a Huson tramway system circa 1901 at the Virginia Chief Mine, Dos Cabezas, Arizona. Note the distinctive towers and automatic loading device below the ore chute. Courtesy Arizona Historical Society, AHS 6955.

Hanson installed a half-mile Huson tramway during the summer of 1891 to avoid some particularly treacherous hauling between mine and mill. Two years later Hanson reported that his $12,000 tramway moved 150 tons of ore per day, worked well, and cost little to operate.[36]

When tycoon Otto Mears began building his narrow-gauge Rio Grande Southern Railroad into the San Juan Mountains in 1891, he opened opportunities to develop additional mining properties, some of which Mears himself invested in. Such was the case of the Black Hawk Mine just east of Rico. Because the railroad could not reach the remote mine entrance, Mears ordered the construction of a Huson tramway during the spring of 1892, remarking that in his extended experience with building mountain railroads in the western mountains he had never seen a better method of "transportation by wire cables." The sixty-five bucket Black Hawk tramway connected the mine with a rail spur. It proved so successful that the mine was soon shipping 150 cars of ore per month, prompting Mears to boast that "the tramway has done excellent work with very little expense in operation, and is thoroughly reliable." Similar other tramways were located along the route of the Rio Grande Southern. As soon as the railroad reached San Bernardo in 1891, for example, a mill and tramway were built. Much the same occurred at the Suffolk Mine at Ophir. The forty-stamp mill located there sported a lengthy aerial tramway, which brought ore down from the mine, passing directly over the town.[37]

The California Wire Works, of course, did not sit idly by while the competition built tramways. During the 1890s, Hallidie and his engineers redesigned the grip pulley and invented a new type of clip for attaching buckets to the cable. The original clip, which used a strap to secure it to the cable, had to be replaced every five months or so. The new clip could be woven into the cable, promising a much longer life and the ability to support a weight of 2,500 pounds "before being sufficiently distorted to prevent its passing the supporting sheaves and horizontal terminal sheaves of the tramway." Unlike the Huson installations, which generally seemed to be less than a mile in length, Hallidie continued to test the reach and carrying capabilities of his apparatus. During the 1890s he built several long tramways in Mexico. One, at the San Juan mines in Baja California, stretched two and a half miles, eliminating the need for a fifteen-mile mule trek. Another tramway near Guadalajara featured a clear (unsupported) span of 2,267 feet. Both of these tramways, as well as one constructed at Monte Cristo in the state of Washington, were noted for their especially steep grades. All of these devices were erected under the personal supervision of E. I. Parsons, the longtime constructing engineer for Hallidie's firm.[38]

Perhaps the crowning achievement for Hallidie's design came in 1895 with the construction of a four-and-a-half-mile tramway at the Silver King Mine near Nelson, British Columbia. All materials, including over 50,000 feet of "special steel cable," were supplied from San Francisco. To build this mammoth system a right-of-way 200 feet wide had to be cleared of heavy timber to make way for the 126 intermediate towers. In order to regulate its speed, supervising engineer Parsons installed "patent grip pulleys, ten feet in diameter, to which are attached strong brakes, to control the speed" of the cable. In all, the tramway descended a total of 4,100 feet using 900 rectangular ore buckets (capacity of 100 pounds each) to deliver silver and copper ore to the smelter at a rate of 10 tons per hour. The cost of carrying ore over the system was estimated at thirty to thirty-five cents per ton. Nevertheless, the strain on the cable proved so great that the system had to be divided into two sections, separated by a transfer station. Even in this configuration, the Silver King qualified as "the largest Hallidie system in the world."[39]

Indeed, the year 1895 was a good one for the California Wire Works, which built five other tramways that summer, all on the West Coast. Nonetheless, the single-rope tramway had reached its limit and would soon be declared obsolete by the mining industry. Hallidie's company continued to defend its product and as late as 1901 was still erecting single-rope tramways. But the heart of the company seemed to fade away when its founder died in March 1900, at age sixty-four.[40] Thereafter, with the leadership gone and the old patents increasingly out of place as mine technology entered the twentieth century, business quickly fell off. Nonetheless, for almost thirty years the single-rope tramways of Hallidie and Huson compiled an impressive record. Perhaps a hundred such devices were erected at locations ranging from Juneau, Alaska, to southern Arizona. There is little question that these machines made it possible for miners to make a profit at locations where traditional means of transportation were impractical. These pioneers, Andrew Hallidie and Charles Huson, paved the way for even greater use of aerial tramways.

CHAPTER TWO
BLEICHERT'S DOUBLE-ROPE SYSTEM

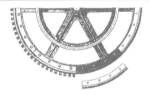

During the 1903 American Mining Congress meeting at Deadwood, South Dakota, Stephen de Zomdoria addressed his fellow mining engineers on the subject of aerial tramways. Praising all types of tramways, he noted that "every mining man of the West can probably cite examples of mines whose profitable operation without the aid of tramways would not be possible." Looking back over two dramatic decades of change in tramway technology following the introduction of the double-rope system, he still found some value in the single-rope tramways of Hallidie and Huson, which continued to be used at western mine sites. They were simple machines, easy to maintain, and relatively cheap to erect and operate. Compared to other forms of transportation, such as roads and railroads, they proved highly adaptable, uncomplicated in construction, and could be taken down and erected at new locations with minimal cost and loss of operating time.[1]

Yet these words offered faint praise. As Zomdoria and others well knew, the single-rope tramway suffered from serious limitations in an era when mine owners demanded more efficiency. One of the most notorious drawbacks involved capacity. Bucket loads had to be limited to less than 200 pounds because anything greater put too much strain on the rope. Indeed, on a device where the moving rope also carried the load, wear on moving parts and constant repairs were inevitable. By necessity, single-rope tramways also

29

ran at a relatively slow speed in order to prevent the rope from jumping out of the sheaves, a problem that continually plagued the Hallidie design. Another inefficiency concerned the fact that buckets were permanently attached to the cable. This "gives rise to the . . . greater objection that the buckets must be both loaded and unloaded while moving, since they cannot be stopped without stopping the whole line." Finally, the single-rope tramways were limited in the distance they could cover, operating with a practical length of about two miles and a maximum of around four miles.[2]

Zomdoria clearly favored the double-rope (or bi-cable) tramway, which by 1890 was generally called the Bleichert system. Perfected by two German engineers, Theodore Otto and Adolf Bleichert, the design improved upon the experiments of others, including Charles Hodgson. Their first successful tramway went into operation in 1874. The double-rope tramway employed a different and more rugged technology. The two-wheeled ore cars (or carriers) ran on heavy stationary cables clamped in place and stretched between terminals to form a track over which the car rolled. Motive power was provided by a second, lighter rope (the traction cable) that ran in an endless closed loop. The traction cable attached itself to the car by means of a friction grip. As described in *The Mining Engineer's Handbook:* "In operation, each load is placed in a carrier standing on a track in the loading terminal; the carrier is then fastened to the traction rope, and is hauled along the track cable to [the] discharge terminal; there it is released from the traction rope and [the] contents are discharged." Major advantages included a higher carrying capacity (up to 2,000 pounds per car), an ability to operate over greater distances, less wear and tear on the cable, and the capability of detaching the ore cars at each end for easy loading and unloading. Although much more expensive to construct, the Bleichert system could move ore at about half the cost of a Hallidie system.[3]

The arrival of the Bleichert design in the United States sparked a contest between advocates of each system that ended with the double-rope becoming dominant by the turn of the century. Nonetheless, it took some time for the German system to become available to western mine operators. The team of Otto and Bleichert built several tramways in Europe before splitting up in 1876, with each continuing in the business independently while continuing to improve the technology. In 1878, for example, Adolf Bleichert & Company built a tramway at the Lintorf Mines and Smelting Works to convey ores from a mine entrance to the railroad and to return coal in the opposite direction. This system reportedly worked so flawlessly that "not a single complaint has been made by the neighbors or the public in general, although the line passes over three public highways. Not a trace of the materials transported can be found on the ground along the line, thus proving that all fears of damage to property in this respect were groundless." By 1880 the success of double-rope systems in Europe attracted the attention of American

The standard design of a Trenton (Bleichert patent) tramway is seen in this 1914 catalog view. The tramway pictured was located at Mound House, Nevada. Author's collection.

TRUCK

FRICTION GRIP

HANGER

LATCH

WILLIAMSPORT

BUCKET

WILLIAMSPORT STANDARD CARRIER

Basic carriage system of the double-rope or bi-cable tramway as illustrated in the 1929 Williamsport Company catalog. Courtesy Williamsport Wire Works, Inc./ Ed Hunter Collection.

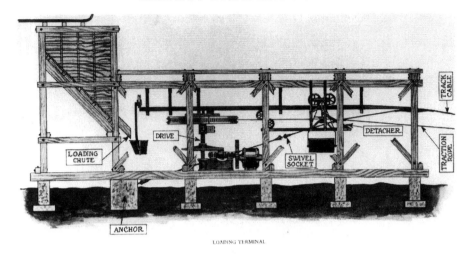

LOADING TERMINAL

This illustration in the 1929 Williamsport Company catalog shows the basic design on the double-rope system loading terminal. Courtesy Williamsport Wire Works, Inc./ Ed Hunter Collection.

mining journals, which noted their superiority over the single-rope system and speculated that "there must be such a substantial advantage in the wire-rod road as to make the question worthy of attention to those interested in the matter." About that time a representative of Bleichert & Company arrived in New York to publicize their invention.[4]

Nevertheless, it took some time for the Bleichert system to be offered for sale in the United States. Not until 1888 was his technology patented in America. Soon thereafter, the Bleichert patents were acquired by the nation's first builder of aerial tramways, Cooper, Hewitt & Company, who operated the Trenton Iron Company. As sole agents for the Bleichert system, Cooper, Hewitt & Company published a pamphlet-type catalog describing the device to potential customers. Using illustrations of some of the more impressive European installations, particularly the seven-mile "Weilburg" tramway, which featured a 1,000-foot span over the Weinbach River Valley, the pamphlet touted the superiority of the Bleichert patent over single-rope designs. According to the text, the new design accepted individual loads up to 2,000 pounds; ore cars safely moved at speeds up to five miles per hour and could be switched from the line at any convenient point. Ropes of different sizes might be used as needed:

> The rope for the empty cars does not of course require to be as strong as the rope for the loaded cars, and is therefore made only strong enough for the work it has to perform. In like manner, if one or more long spans

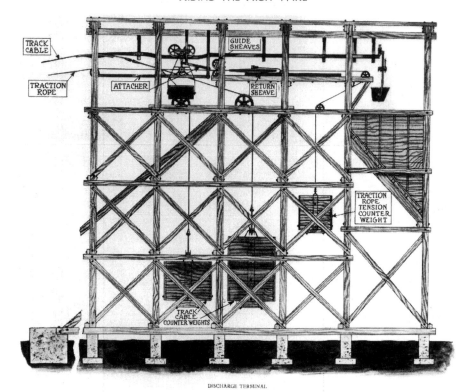

DISCHARGE TERMINAL.

This drawing from the 1929 Williamsport Company catalog diagrams a typical double-rope discharge terminal. Courtesy Williamsport Wire Works, Inc./ Ed Hunter Collection.

occur in the line, it is not necessary that the whole rope should be made strong enough to bear the extra strain at this one point; on the contrary, it is sufficient to so strengthen only the portions exposed to the extra strain. By means of our patent couplings this is easily practicable.[5]

The Bleichert offered other advantages, especially with the interchangeability of carriers that could be used to haul all sorts of material. Indeed, the 1888 catalog included cars designed to carry lumber and pipe, boxes or bales, cordwood, barrels, and ores. To bolster claims of versatility, the company listed some fifty-two European tramways it had contracted for or constructed in 1886. Varying in length from a few feet to over three miles, they serviced mines, brick factories, chemical works, weaving mills, cement works, and quarries. Although not directly focused on the western mining industry, readers of the descriptive catalog were invited to write the engineering department at Cooper, Hewitt & Company in New York City to obtain cost estimates.[6]

Despite the then dominant hold of single-rope tramways on the western market, the Trenton Iron Company, which manufactured both the rope and the machinery, advertised in mining journals with gratifying results. Within a few months over two hundred companies had asked for particulars on the Bleichert tramway and several signed construction contracts. Although eastern coal mines and quarries provided the majority of the early orders, the lure of western profits saw the Trenton Company entering the Rocky Mountain region just as the boom of the early 1890s exploded. In 1889–1890 the Trenton works completed the construction on two silver-ore tramways at Aspen, Colorado. The Compromise Mining Company line was barely more than a half mile in length, but the Aspen Public Tramway Company system reached almost two miles, with a fall of 2,409 feet. Each line carried 250 tons of ore per ten-hour shift. Soon thereafter, the Trenton Company set up a 6,240-foot tramway for the Silver Age Mining and Milling Company at Idaho Springs, Colorado.[7]

Despite many positive comments about these early Bleichert systems, things did not always go well. Nowhere was this more evident than at the Bunker Hill and Sullivan Mine at Wardner, Idaho. Located in the Coeur d'Alene Mining District, this rich silver-lead mine dated back to 1885, although significant production did not begin until 1890. In 1891 the Bunker Hill built a new concentrator at Kellogg and contracted with the Trenton Iron Company to construct a 9,000-foot tramway between the mine and mill. The line ran directly over the town of Wardner, where a tension station housing a twenty-ton weight was located. The track cables were probably of the newly patented interlocking design, which featured "outer wires of such shape that they interlock with each other, . . . presenting a smooth surface, and yet possessing sufficient flexibility to be shipped in coils. This cable gives the highest degree of service with minimum wear on the carriage wheels."[8]

Troubles began soon after the line entered service. About five in the afternoon of September 23, 1891, residents of Wardner were jolted by a sharp crack as the traction cable snapped when the heavy tension weight broke loose. One end of the cable whiplashed around the town, taking down utility poles and crashing into several houses. Several ore buckets also toppled onto the homes below, killing a woman. The remaining buckets, set free by the cable break, started rolling down the track toward the concentrator. "Buckets came down in rapid succession," noted the *Spokane Spokesman*, "their arrival resembling the pattering of rain. A shed was demolished and hundreds of buckets were deposited in wild confusion." In all, the disaster cost the company about $25,000 and forced a month-long work stoppage. No sooner were repairs completed that the line broke again. This time the Bunker Hill management insisted that the Trenton Company replace the traction cable, which they did in November.[9]

35

View of the Bunker Hill and Sullivan tramway at Wardner, Idaho, after being rebuilt in 1892. Courtesy Trenton Iron Company catalog, 1896/ Ed Hunter Collection.

Nevertheless, the tramway continued to malfunction. In January 1892, mine manager V. M. Clement wrote company president John Hays Hammond to express his frustration. Addressing "the shortcomings of the Tramway," he argued that "as a matter of right the Trenton Company should place at their own expense, this tramway in satisfactory working condition for 400 tons of ore & 50 tons of upfreight per 10 hours as per contract." Clement believed that the tramway could be made acceptable, but only with radical changes, including replacing all the carriers. So far, Trenton had failed to live up to its obligations despite their claim that everything was "working to full satisfaction." As the mine manager noted, "if such is their conception of a perfect working contrivance, then the less dealing we have with them the better off we will be." If the manufacturer refused to make things right, Clement proposed spending no more money and instead begin building a railroad connection.[10]

Trenton officials, apparently stung by the complaints, agreed in the spring of 1892 to make major changes in the Bunker Hill tramway at their own expense. Basically stated, they removed the tension station at Wardner and two of the towers, giving the line a clear span of 1,173 feet as it passed over the town, thus eliminating "the difficulty experienced in transferring the cars, on account of the steep angles of the converging cables." When the repairs were completed, the 9,000-foot line featured two intermediate tension stations to offer additional support as it passed over the crest of the mountains. At these sites "the cars pass from one section of the cable to the next by means of intervening rails, so that no interruption occurs in the continuity of the track." Because of frequent storms, these stations were protected with snow sheds. Bunker Hill officials wanted to test the rebuilt line in May, but owing to a bitter strike then in progress, they could not secure enough ore until a month later. At that point, the Trenton Company turned the tramway over to the company for a second time.[11]

Even these repairs failed to resolve all the problems. During the month of July several more accidents occurred. Buckets continued to fall off the line and at one point a tower was pulled down. Continued tinkering, however, finally produced improvements. During August 1892 the line ran a total of 258.5 hours with only 21.5 hours of lost time, none attributed directly to accidents. Yet the carriers still seemed so unreliable that the following month Clement entered into an agreement with the Trenton Iron Company to furnish him 160 new carriers at half price. This seemed to work and by December assistant manager F. W. Bradley could write Hammond that "the tramway continues to do good and satisfactory work—it has run so far this month without accident."[12] From this time onward, the Bunker Hill tramway continued to have a checkered career, experiencing various operational problems. As a consequence, it was retired in 1902 with the completion of the Kellogg Tunnel.

During the early 1890s the Trenton Iron Company constructed about twenty aerial tramways in the mining west, some of which attracted considerable attention for their innovation. The bucket line built for the Old Dominion Copper Mine at Globe, Arizona, in 1891–1892 illustrated that a system did not have to be long and steep to provide significant savings. In this case, the distance between mine and smelter was only 1,224 feet, with a drop of barely 100 feet. Nonetheless, a deep gully separated the two points, meaning that ore had to be hauled down a steep road by mule team. As at many other locations, moving ore wagons up and down such grades proved both dangerous and costly. On the downhill trek mule skinners had to take care that the heavy ore wagon did not break away, run over the mules, or slide off the road. The return journey uphill to the mine was nearly as difficult, it being almost impossible to move the wagons up a steep trail. In addition, the mules had to be fed and watered. Consequently, transportation of ore at the Old Dominion cost about twenty cents per ton. When the new tramway entered service in January 1892, it utilized eleven buckets of 800 pounds capacity each, which enabled the company to deliver by gravity as much as 170 tons of ore per ten-hour shift. As *The Arizona Silver Belt* concluded, "the advantages of the tramway over hauling the ore in wagons are important. The tramway can be run in all kinds of weather, and the cost of delivering ore in the bins is 9 cents per ton, against 20 cents by wagon."[13]

Other Trenton tramways were more spectacular. Although they could be found from Juneau, Alaska, to Kelly Switch, New Mexico, the largest number operated in Colorado. The early 1890s silver boom at Creede produced such a flurry of activity that three tramways went up in record time, helping the town become Colorado's leading silver producer. Almost simultaneously, the Holy Moses and Amethyst mines ordered bucket lines. Although only 2,000 feet long, the Holy Moses tramway was considered to be the steepest one in the United States. With a drop of 1,020 feet, portions of the line sat on a thirty-one-degree incline. In its own way, the Amethyst was equally spectacular. Over 8,250 feet long as it reached into the rugged mountainside, it required the construction of several sixty-foot wooden towers. One historic photograph shows workers perched precariously on an uncompleted tower during construction in the fall of 1892. The Bachelor-Commodore Mine tramway featured an unusual arrangement, in that a short tram line brought ore down from an upper mine entrance to a lower one. Here the load was discharged into a bin, then loaded on a second line for the longer trip to the discharge terminal.[14] How long these lines remained in operation is unclear, as the 1893 drop in silver prices produced a significant decline in mining activity.

The Trenton Iron Company was so pleased with its product that it volunteered to construct an expensive demonstration tramway at Chicago's 1893 Columbian Exposition. The 1,000-foot Bleichert system ran from an ore storage yard, up and over the Intermural Railway, to a discharge terminal attached to the southwestern corner of the Mining Building. Powered by electricity and utilizing steel towers, the exhibit presented a dramatic sight for fair visitors as they strolled along a path paralleling the line. Aside from illustrating a piece of technology, the tramway actually served a purpose. As one account noted: "These buckets come and go with the persistent regularity of clockwork, and what adds interest to their passing is the fact that early in the day they transfer diamondiferous ground from the store yard at the distant terminal to the Mining building, whence it is transferred to the crushers which operate daily in the South African diamond exhibit in this building." The Trenton Company also provided an indoor exhibit displaying various kinds of wire rope and mining cars.[15]

Although the Bleichert tramway exhibit at the World's Fair attracted the greatest attention, it was not the only one there. Both Hallidie's California Wire Works and C. W. Badgley (Huson system) offered examples of the single-rope system, each touting its specific features. Even more interesting, J. Pohlig Company of Cologne, Germany, contributed a "full size exhibit of the Otto system of wire tramway." Using the designs of Theodore Otto, Adolf Bleichert's former partner, Pohlig's double-rope tramway in the Transportation Building was nearly identical to the Trenton offering. Not yet marketed in the United States, it nevertheless seemed clear that Pohlig intended to compete with his American rivals.[16]

The discharge terminal of Trenton's exhibition tramway at Chicago's World Fair in 1893 as depicted in the company's 1896 catalog. Courtesy Ed Hunter Collection.

Because of the growing popularity of aerial tramways and the continuing need for transportation at western mines, additional companies inevitably entered the market. The 1890s and early 1900s thus saw a proliferation in tramway-manufacturing companies. With few exceptions they used technological variations of the single- or double-rope design.

Surprisingly, perhaps, a number of companies introduced upgraded versions of the single-rope system. In its 1896 catalog the Trenton Iron Company offered a single-rope line called the "Roe System." Its design promised to overcome "all the objectionable features of the Huson and Hallidie" by making it possible to carry larger loads and thus produce greater outputs. It also featured simple construction and an ability to climb steeper grades at higher speeds. As early as 1891 the Vulcan Iron Works of San Francisco began offering a "patented" single-rope tramway that closely resembled the Huson type. Somewhat more sophisticated was a device introduced in 1900 by the J. H. Montgomery Machinery Company of Denver. This technology permitted the bucket, when arriving at the loading terminal, to be released from the cable for easier loading. As one report noted, "Any one who has ever used a single-rope system will understand the great advantage of having the buckets stop to be loaded, as the ore is not scattered all over the station, and much complicated machinery, which must be kept in repair, is dispensed with." The ore buckets, which were turned upside down to empty, offered the additional advantage of being returned to the mine in that position during stormy weather to avoid being filled with rain or snow.[17] Despite such improvements, it does not appear that the new single-rope varieties ever became particularly popular with western miners.

With double-rope systems clearly the wave of the future, many more companies went in that direction, each with its own variation on the basic Bleichert design. In 1892 the Union Wire Rope Tramway Company, a firm closely associated with cable manufacturer John A. Roebling's Sons, completed a tramway across the Penobscott River in Maine. The unique feature here was a newly designed grip that locked the traction rope to the carrier without injuring the cable. Although similar to systems used in the West, it seems doubtful that this company ever made much of an effort to move in that direction. More competition came from the introduction of "Otto Aerial Tramways." By the late 1890s Pohlig began to license this system in the United States, at first through the Chicago firm of Fraser & Chalmers. Noting that some seven hundred Otto tramways had been built in Germany and Europe, by 1898 Fraser & Chalmers had completed lines at the Florida Mountain Mine in Idaho and the Yellow Mountain Gold Mine in Colorado. The selling point for the Otto system also involved the grip. Illustrating the "Pohlig Universal Friction Grip," one article described its advantages thus: "Owing to the fact that the gripping of the rope is not confined to certain spots, but may occur at any point, and that the grip is free to turn on the shaft, thereby allowing the jaws to adjust themselves parallel to the direction of the pull and sag of the hauling rope, the wear on both rope and grip is reduced greatly."[18]

Other companies also joined the field. Vulcan, which earlier marketed a single-rope system, soon offered a double-rope system. Meanwhile, the A. Leschen & Sons Rope Company of St. Louis, which had entered the cable business in 1886, began to "manufacture, buy, sell, build and erect Aerial Wire Rope Tramways and Cable Ways." Leschen soon emerged as one of the major American tramway companies, remaining in business until 1953. Another St. Louis rope manufacturer, Broderick & Bascom, also began to construct mine tramways around the turn of the century. Their apparatus featured the "Desedau" design with a "special patent clip" and guide sheaves that made it possible for the traction cable to remain "at the same distance from the standing cable whether the bucket has passed over the tower or not, thereby increasing the life of the cable." By way of contrast, the traction cable on the Bleichert system sagged between buckets, resting on rollers attached to the towers. As early as 1902, Broderick & Bascom had installed a Dusedau tramway at the Contention Mine near Silverton, Colorado.[19]

By all odds the company to most successfully compete with the Trenton Iron Company was founded by Byron C. Riblet in 1897. According to a story told by Riblet, he became involved in the tramway business by accident. Born in Iowa in 1865, Byron earned a degree in civil engineering from the University of Minnesota in 1886. Hiring on with the Northern Pacific Railway as a surveyor, he eventually landed in Spokane, Washington, where he installed a streetcar system and became chief engineer for the Washington Water Power Company. In 1896 the Noble Five Mining Company of Sandon, British Columbia, hired him to build a water-power plant, and, so he thought, a mine railway. When he reached Sandon he was surprised when one of the contractors said, "I'm glad I got you to do the work, for there are few people [who] know anything about aerial tramways. But with an expert like you, we're all right." At that point Riblet exclaimed, "Aerial tramways, I never saw one." Prepared to return home, he was persuaded to stay on. Accordingly, he consulted as many trade books as he could find and set about building a tramway.[20]

Riblet's double-rope tramway at Sandon proved to be generally successful. Some 6,100 feet in length, with an elevation drop of 2,050 feet and one span of 1,000 feet, it worked reasonably well, although he found numerous defects that could be improved upon. Consequently, Riblet "made improvements and took out patents." Other British Columbia miners were sufficiently impressed with the Noble Five tramway to encourage Riblet to open a fabrication shop in Nelson, British Columbia, under the name of B. C. Riblet, Engineer. He quickly called in his brothers Walter and Royal to run the office and make sales. Byron meanwhile continued to improve his technology, patenting a bucket grip in 1902 and developing "a terminal chute to load moving buckets, a novel scheme in those days." Not yet fully committed to the tramway business, however, Riblet sold some of his patents to the Leschen Company, then joined them as chief engineer, building as many as thirty aerial tramways, including the sixteen-mile line at Grand Encampment, Wyoming (see Chapter 4). In 1908 he returned to Spokane and three years later founded the Riblet Tramway Company with his brother Royal, who had remained in the tramway business during Byron's absence. From here the company developed a business that constructed mine tramways all through the Western Hemisphere, focusing heavily on western Canada and Latin America.[21]

Although the double-rope tramway grew increasingly popular, well into the late 1890s proponents of the single-rope design continued to believe they could be competitive. Perhaps the only head to head comparison between the two technologies occurred during the Klondike gold rush of 1897–1898. Miners headed to the Yukon were faced with crossing either White Pass or Chilkoot Pass north out of Skagway or Dyea, Alaska. Stories of miners hiking over these rugged trails to reach Canada are legend as some hundred thousand people made the long journey to the Dawson goldfields under incredibly brutal conditions. All supplies had to be packed over the mountains, with Canadian authorities insisting that each man bring sufficient provisions to survive for a year in the interior. Most of this human traffic eventually went over the 3,500-foot Chilkoot Pass. Indeed, some of the most famous images of this historic event depict endless lines of "Stampeders" struggling over the pass during the winter of 1897–1898. Because of the great difficulty in moving men and supplies, a number of businessmen began to look toward the construction of aerial transportation over the most difficult parts of the trail.[22]

While the Chilkoot tramways were not technically mine-related—they were built to move freight and even passengers over the pass—they employed the same technology and attracted considerable attention. Three tramway projects began in late 1897 after several attempts to construct horse-drawn sleds and surface trams failed to solve the problem. The Dyea-Klondike Transportation Company (DKT) completed the first line, opening for business on March 14, 1898. Only 2,400 feet in length, the DKT tram attracted attention because it was powered by electricity, having a steam-fired powerhouse at the foot of the tramway. The manufacturer of this simple device is unknown. It consisted of two cables stretched without support between terminals. Two buckets of 500 pounds' capacity rolled on wheels along the cables. This may have been a modified Bleichert system, although it apparently proved unsuccessful, and probably ceased operation during the summer of 1898.[23]

Meanwhile, the Alaska Railroad & Transportation Company (AR&T) decided to construct a single-rope tramway running about a mile and a half up and over the summit. National Park Service surveys of the remaining artifacts suggest that the AR&T used the Huson design. Construction probably began in early 1898 and was completed in May that year. Goods were carried uphill in a series of buckets, the power being provided by a gasoline engine. Although much more efficient that the DKT tramway, the AR&T operated for less than a year. It apparently began losing business as soon as the much longer and more effective Bleichert system was placed in operation by the Chilkoot Railroad & Transport Company (CR&T).[24]

The longest, most successful, and best-known Chilkoot tramway thus belonged to the Chilkoot Railroad & Transport Company. Originating as an investment opportunity during the summer of 1897, the first proposal included a combination of roads, a horse-drawn tram, and a tramway to reach from the port of Dyea to Lake Lindeman, a distance of twenty-six miles. Financial realities soon compelled the company to opt for a nine-mile double-rope tramway manufactured by the Trenton Iron Company. Construction began in December and proceeded well until the great avalanche of April 3, 1898, which killed an estimated twenty-three construction workers. When the line finally opened in May, it sported a number of unique features. William Hewitt of the Trenton Company noted that all materials had to be kept as light as possible. As a consequence, they selected a "smooth coil" track cable only seven-eighths of an inch in diameter. Being made from special crucible steel, it could support 36,000 pounds while weighing less than three-quarters of a pound per foot. This compared favorably to ordinary cable, which ranged from two to five pounds per foot.[25]

Chilkoot Pass tramways circa 1898. Shown in the picture are the Bleichert double-rope tramway (tower at far left) and a single-rope tram (center) carrying loads of freight. Courtesy Alaska State Library, PCA 39-850.

Because of the line's great length, and after a blizzard blew the cable down, it became necessary to apply tension at two intermediate points. As described by Hewitt, "the track cables are parted at these points. . . . The cars pass from one section to the next by means of intervening rails, so that no interruption occurs in the continuity of the track." The Chilkoot tramway, its promoters incorrectly proclaimed, was also the first one designed to provide for passengers as well as freight. For supplies, the system relied on buckets capable of holding between 200 and 300 pounds each. Passengers generally rode on a boatswain's chair type of conveyance. Although the CR&T expected to carry a goodly number of people over the treacherous mountain, it is unknown how many actually rode the line. At least one woman, however, did cross "in a hour and a half the mountain defile which has hitherto tried men's souls and bodies in a struggle of days and weeks." This passenger, Martha A. Kelsey, had been strapped into a freight box to become the "first woman to be carried over the new aerial tramway." The trip may not have been glamorous, but it proved effective. As Hewitt noted, "It is needless to say that all will travel alike; there will be no drawing room cars."[26]

The CR&T tramway proved successful, providing an economical and efficient way to lift materials over the pass. Freight rates dropped to such a degree that Indian packers were put out of business and toll road operators were forced to lower rates. During the summer of 1898 the C&RT reorganized itself into the Chilkoot Pass Route by buying out the two smaller tramways, which were either shut down or had already ceased operation. Despite its efficiency, the surviving system soon found itself faced with a competition it could not defeat when the White Pass & Yukon (WP&Y) narrow-gauge railroad began construction in 1898. The tramway remained in service, offering competitive rates, until the WP&Y reached Lake Bennett in 1900. At that point, the railroad purchased the tramway operation and shut it down as a means of eliminating competition.[27]

The success of the Bleichert design on Chilkoot Pass symbolically demonstrated its superiority over other systems. T. A. Rickard, after visiting various tramway installations in the vicinity of Silverton, Colorado, in 1903, observed that the contest was over. Noting that most current installations were of the Bleichert or Otto type, he declared that the Hallidie and Huson single-rope systems were "nearly obsolete except for short distances and over easy contours." Experience, he stated, "now favors the double ropeway system in spite of a cost of installation which is 30 to 50 per cent greater than the single rope type, because this difference in first cost is soon wiped out by the cost of maintenance, which with the Hallidie type is nearly double that demanded by the Bleichert." Additionally, the double-rope, because of its heavy construction, could move up to four times the amount of ore.[28]

Rickard's comments, which appeared in the *Engineering and Mining Journal*, drew a challenge from Arthur Painter of San Francisco, who apparently represented the Vulcan Iron Works. Agreeing that the Hallidie and Huson systems were indeed obsolete, he claimed to have recently installed a five-mile single-rope line so efficient that it delivered twenty tons of ore per hour at a cost of eighteen cents per ton, including labor and repairs. These claims were quickly challenged by R. D. Seymour, who responded that if Painter's comments were in fact describing a tramway recently constructed in British Columbia, then "he is certainly in error claiming any such economical cost of operation." Taking the Camp Bird tramway near Ouray, Colorado, for comparative purposes, Seymour stressed that Painter failed to include such factors as back traffic and reliability in his calculations. Seymour thus agreed with Rickard: "The single rope system of aerial tramways has served its time and is certainly out of date now. Very few mining engineers would consider using it except as a cheap installation to do prospecting or other temporary work." Most other mining engineers seemed to agree with this conclusion.[29]

The advocates of the double-rope system proved to be good forecasters. While tramways of the Bleichert model continued to be expanded and improved, the single-rope fell into oblivion. To be sure, the Trenton Iron Company and other manufacturers continued to offer inexpensive single-rope systems through their catalogs, but they attracted little interest. Perhaps the best sign of the demise of this design came in December 1905, when the California Wire Works (Hallidie's company) went out of business and forfeited its charter for failing to pay the state license tax.[30] An era had truly ended.

Literally hundreds of double-rope tramways went up in the mining West between 1900 and 1920. In some locations they filled the sky, with hundreds of buckets quietly moving millions of tons of ore daily along miles of steel wire track. For the most part, Leschen, Riblet, and Trenton dominated the business, although a number of smaller companies also contracted to install systems. The Trenton Company, probably the most prolific of all the fabricators and still the sole American licensee of the Bleichert patents, came under control of the United States Steel Corporation in July 1904, although the marketing department continued to use the Trenton name. By 1910 the Trenton tramway factory was listed in catalogs as a division of American Steel and Wire Company.[31]

So important were these tramways to successful mineral production that an anonymous engineer penned this tribute just after the turn of the century:

> Nestled silently in the clouds, away up above the timber line, nature has hidden almost unaccessibly [sic] its treasures. . . . Reluctant to give up her treasures, nature wages a constant warfare with man, calling to her aid the snows and ice of winter, altitude, precipice and ravine. But man has won the fight. Two slender wire cables, puny in appearance despite their strength, span ravines, rise over precipices and scale the heights, disappearing among the clouds. . . . Silently, unpretentiously, disdainfully ignoring the grumbling of nature over her defeat and her efforts to overthrow the work of man, the cables modestly move forward, the connecting link between mountain and valley, a private soldier of the mining industry, always alert, always performing his duty, always his commands and rendering invaluable service in adding to the wealth of the nation.[32]

CHAPTER THREE
CONSTRUCTION AND OPERATION

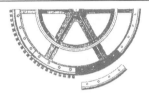

Wherever they were built and whatever the design, aerial mine tramways in the western regions of North America were constructed and operated with a certain degree of uniformity. They were, after all, built for the economical conveyance of mine products over distances ranging from a few hundred feet to many miles. Despite the similarities, each tramway was unique, covering different topography and presenting a variety of operational and construction challenges. This chapter will examine in general how the tramways were designed and built, problems and maintenance, accidents, economic and labor factors, and the dangers of passenger traffic.

The first task of any mine operator involved an evaluation of the cost-effectiveness of a tramway. Decisions were usually based on the distance between reduction facilities or rail transportation and the mine entrance. A rule of thumb used by many mining engineers said that double-rope tramways of less than a thousand feet were impractical because terminal machinery and labor cost as much as that of a longer line, making initial costs prohibitive. Thus where short distances and level terrain were involved, it might be more practical to use mule trains or other forms of surface transportation. Once great distances and rugged terrain entered the picture, however, miners gave serious consideration to erecting an aerial tramway. Although the initial cost might be considerable, in most cases tramways were cheaper than such alternatives as constructing narrow-gauge railways. In 1903, T. A. Rickard, then editor of the *Engineering and Mining Journal*, pointed out the advantages of choosing a valley mill site "and transporting the product of the mine over an aerial tramway. . . . As a rule the valley site is preferable, because of the availability of a water supply, the greater cheapness of fuel for power and heating purposes, the nearness to a base of supplies, the facility which the tramway itself gives for transmitting materials up to the mines, the more kindly conditions of living for workmen, etc." Substantial tramways also served to convince investors that the mine in question was no fly-by-night affair. While this assumption was not necessarily valid, an aerial tramway on the property, along with other visible signs of prosperity, no doubt helped to sell stock. Reports to shareholders and potential investors frequently touted state-of-the-art machinery. The 1899 general report on the Silver Lake mines near Silverton, Colorado, for example, not only fully described the current tramway system, but also noted that additional mineral resources would be "gradually exploited as the facilities for handling the ore shall be increased, through the building of further Tramways, Mills, etc."[1]

Once having made a commitment to use tramway technology, the mine owner faced several options. He could utilize his own employees or private consulting engineers to design the line, purchase the materials, and put it up. Another, much more common alternative involved contracting with one of the manufacturing companies to use their engineering staff to design and build the line. The appeal of the latter type of arrangement seems to have centered on the fact that the manufacturer would stand behind his work. Such agreements appear to have become common about the time the Trenton Iron Company introduced the Bleichert system in the late 1880s. Certainly, when Idaho's Bunker Hill tramway experienced a rash of problems in 1891–1892, it expected the Trenton Company to make it work properly "as a matter of right," if not an absolute guarantee. Trenton's management honored this assumption.[2] In another case, Leschen & Sons offered to construct a short tramway for the Bull Hill Mining and Development Company in 1906. Leschen agreed to furnish everything from the cable, tower structures, and end terminals to bolts, rods, and washers for a package price of $1,775. To facilitate construction and make adjustments, the company also offered to furnish "a thoroughly competent superintendent" for $8.00 per day plus expenses. If erected under the direct supervision of this man, Leschen agreed that the tramway would satisfactorily transport ten tons of ore per hour.[3]

Before construction could begin the route had to be surveyed and a right-of-way acquired. In areas of difficult terrain the cost of surveying might be considerable, particularly if months of preliminary work were necessary to discover the most economical route. While tramways were generally laid out in a straight line (since this represented the shortest distance between terminals, and thus the cheapest to construct and operate), it was occasionally better to deviate from the shortest route. This was especially true where a tramway passed over a sharp summit, making it more advantageous to angle across a lower pass. Once the route had been surveyed, it became necessary to secure permission to build the line across public and private property. In many western mining regions, land values were negligible and permission to pass over property could be easily obtained. Compared to other forms of transportation, tramways caused relatively little environmental damage and required only a few hundred square feet of ground per tower.[4] In other instances, however, securing a right-of-way presented significant financial and legal problems.

Some companies preferred to acquire the right-of-way by outright purchase. This was the case on the Bunker Hill and Sullivan tramway. In 1892 its board of directors opted to purchase the property below their tramway to forestall litigation after several accidents caused property damage. Colorado's Aspen Mining and Smelting Company took a different approach. In 1887 it laid out a tramline to transfer ores from its mine to the "Hewitt Sampler" in the town of Aspen. About two hundred feet of the route passed over the claims of three individuals, who refused to grant a right-of-way. Arguing that state laws permitted it to condemn privately held lands for tramway purposes in the same way that land could be condemned to construct a public highway, the company filed a petition to that effect in Pitkin County Court, deposited $200 to compensate the landowners, and proceeded with construction. The three landowners objected and secured a court ruling in their favor, whereupon the company appealed to a higher court. Eventually the state supreme court ruled that tramways did not have "the right to appropriate private property for private ways of necessity" in Colorado. The Aspen Mining and Smelting Company thus had to make a separate and more costly settlement with the landowners.[5]

Laws in other states differed, however. The Hercules Mining Company of Burke, Idaho, sought legal advice in 1904 about acquiring "lands belonging to other persons and corporations without their consent." The issue also involved taking land for "public use." M. S. Folsom, the company's attorney, stressed to his clients that any such taking would have to comply with the due process clauses of the United States and Idaho constitutions. The Idaho document seemed relatively favorable, stipulating that private property could be "taken for public use" after paying just compensation. Mining activities, including roads, railroads, and tramways, regarded as necessary to the state's development of mineral resources, thus came under the constitution's public use provisions. The statutes of Idaho, Folsom wrote, were therefore sufficiently clear "to authorize you to proceed to condemn a right of way for railroad or a tramway" providing it did not violate the constitutional rights of the property owners. Thus, mine owners would probably have to prove in court that the taking constituted a public use. As the attorney noted, the supreme courts of Montana and Nevada had ruled that mining constituted a public use, while other state courts (like Colorado's) rejected the argument. Folsom believed, however, that Idaho courts would support the taking, although "I would not feel justified in advising you that it would [be] an absolute certainty."[6]

An additional right-of-way cost involved clearing trees in heavily timbered areas. To prevent trees from interfering with operations, a strip of land fifty feet on each side of the center line would ordinarily be cleared. In areas of unusually tall timber, the strip might be even wider to avoid the possibility of a falling tree damaging the line. Such clearings could easily cost up to $2,000 per mile and needed to be accomplished before construction could began.[7]

Constructing a tramway was often difficult and dangerous. Workers were required to erect high towers, angle stations, rail stations, tension stations, and terminals in addition to stringing the cables. Much of this was done in rugged mountains, with construction crews forced to scale sheer rock faces to erect the necessary structures, in some cases suspended by ropes. Designers preferred timber as a construction material because it was sturdy and could be locally obtained. Such woods as Douglas fir and ponderosa pine stood up well and often lasted longer than the tramway itself. Contractors generally used 6" x 8" or 8" x 8" timbering for towers, stations, and terminals, making these structures quite substantial. Although more expensive, steel towers were used in a few locations. Stronger than wood, it could be prefabricated, and required less maintenance. Because it had to be ordered from an eastern manufacturing plant and transported over great distances, however, overall costs were much higher. Not until well into the twentieth century did steel towers become more commonplace (although they are known to have existed in the 1890s). By the late 1920s the Riblet Company had developed a small portable fabricating plant that could be used in the field to produce steel components, which reportedly made them as cheap as timber.[8]

Towers, either wooden or steel, came in a variety of designs. They needed to be strong enough to withstand the downward pressures created by the weight of the cable and carriers, have the ability to hold up to changes in the direction of pressure as the load approached and departed the tower, and to stand up against high winds. Many single-rope tramway towers were of relatively skimpy construction, reflecting the light loads carried by such systems. Double-rope tramways, which were capable of carrying up to a ton of ore per bucket, needed to be more substantial. Several basic designs were commonly used on the double-ropes. One of the most popular, the pyramid type, angled upward from a square base to form a pyramid shape. Attached to the top, which might be as high as a hundred feet above the ground, were saddle beams holding the track cables. Below them a roller deck fitted with sheaves or brackets held the traction cable when slack. Another common design was

This photograph of a massive tramway tower under construction in the Colorado Rockies gives some idea of the rugged terrain that designers had to contend with. Courtesy Colorado Historical Society, F28.429.

An example of the daring construction necessary to build a cable line and towers is this scene of the Sunnyside tramway looking up Eureka Gulch near Silverton, Colorado. Courtesy San Juan County Historical Society Collections.

called a "through tower," regarded as strong and efficient, but requiring twice as much timber as the pyramid. On this type of tower, a substantial braced frame supported the roller deck. Above that, the track cables were secured to the saddle beams attached to a Y-type support and enclosed with timber framing. Most other types of towers were variants of these two designs. Engineers preferred to have towers spaced every 200 to 400 feet, but much longer spans were often required. Some designers considered extremely long spans to be uneconomical, although others found no objection to them when necessary.[9]

Terminals and various on-line structures rounded out a typical tramway system. For systems extending for more than a mile or two, it was often necessary to divide it into sections to prevent excessive cable sag. Section stations transported a carrier from one section to another by means of a rail connection. These structures were also used to add tension to the line in the form of heavy counterweights attached to the end of the cable. Terminals came in a variety of forms. A discharge terminal (essentially a holding bin for ore) could be constructed at a mill site or at a place suitable for transferring ore to other forms of transportation. Most of the latter type of terminals were built for the purpose of loading railroad cars. At a few locations—such as the Mountain Hero Mine at Conrad, Yukon Territory—the discharge terminal was constructed on a wharf for loading into barges. During the open-water season, small ore shipments were forwarded to Carcross and then on to Tacoma, Washington, for processing. Many discharge terminals were quite substantial. At the Christmas Copper mine in Gila County, Arizona, for example, the lower terminal bins held over 1,000 tons of ore. The ore bins at the Silver Lake Mill near Silverton, Colorado, were even larger, holding 1,200 tons.[10]

A final task involved selecting and installing the cables. Well before the turn of the century, wire rope manufacturers were producing many different types of steel cable. Two basic types of track cable were usually available—the spiral (or smooth-coil) and the locked-coil. The spiral type (coiled strands of individual wires similar to hemp ropes) had the advantage of lower cost, but was much less capable of transporting heavy tonnage. Because it did not present a perfectly smooth surface, it was also more susceptible to wear and breakage as well as being more likely to damage carrier wheels. Another disadvantage involved the tendency for moisture to penetrate between the individual strands and cause rust. For these reasons, locked-coil cables were generally preferred, especially on the longer systems. As described in a Riblet catalog: "The Locked Coil Cables . . . are of Cast Steel and of a construction that gives a maximum wearing efficiency, both to the Cable itself as well as the Truck Wheels which run on it. The outer covering of Locked Coil Cables consist[s] of specially shaped wires which interlock and produce a smooth unbroken surface." Both types of cables could be spliced together with special couplings that were tapered to fit the cable, thus permitting easy passage of carriers. Traction ropes were smaller cast-steel spirals of six or seven strands.[11]

Moving the cables to the construction site and putting them in place proved a particularly difficult and dangerous task. To avoid splicing, the cables

Downtown Telluride, Colorado, April 1897, with sixty-five mules loaded with 15,000 pounds of cable destined for the Nellie Mine tramway. Courtesy Denver Public Library, Western History Collection, X-152.

came from the wire factories on giant spools, some holding as much as 16,000 feet of wire rope and weighing over 15,000 pounds. When the Bunker Hill and Sullivan tramway received a replacement cable from Trenton in 1891, the heavy load had to be placed on "a specially strong wagon and drawn by nine teams to the power house."[12] In mountainous areas where roads were unavailable, cable might be moved to the construction site by pack train. A packer in Telluride, Colorado, who in 1897 undertook to move 15,000 pounds of cable to the Nellie Mine tramway on the backs of sixty-five mules, claimed to have invented this technique. A description of this impressive process appeared in the August 7, 1897, issue of the *Mining and Scientific Press*:

> The first step was to arrange the cable for packing, and this was done by unwinding from the spool and recoiling it into 130 small coils of some 3 or 4 feet in diameter. These coils were tied with wire and laid about 9 feet apart. When this was done the mules were brought down from the barns at Pandora and strung along the coiled cable that stretched out over two blocks of ground. Beginning with the lead mule the packers then raised

This drawing from the 1910 Trenton Company catalog details the way rope was loaded and transported by mule train. Author's collection.

each set of two of the small coils, placed them over the saddle, to which they were firmly lashed so as not to shift from an even balance on the trail. The work of loading the cable was completed in less than three hours, and after posing for the photographer the train of sixty-five mules with its load started for its destination on the tramway line.[13]

Much the same technique was used at the Silver Dollar Mine near Camborne, British Columbia, in 1907. In this case, a rough and steep mountain trail presented so many obstacles that the mill machinery barely survived the trip. To bring up the cable, a local freighting firm divided up the traction and track cables into seven sections, then loaded each section on horses for the six-mile trip. Each day for a week an increasingly larger section of cable went to one of the terminals. On the last day, some "31 horses conveyed 3600 ft. of $1\frac{1}{8}$ in. cable, weighing 7600 lb., to the upper terminal. All the cables were delivered in good order, without a kink in them."[14]

Under ideal conditions the cable arrived at the terminal site still on the wooden reel. There the reel would be mounted on a steel axle and raised off the ground. The cable could then be played out along the right-of-way, being careful not to cause a kink. Where the line crossed a road or other obstacle, it would be raised on wooden supports. Splices were also made at this time. Track cables were then lifted to the top of each tower with a block and tackle and secured in place. After this task was completed, the weights were connected in order to bring the line under proper tension. Another method of stringing the track cable, particularly in mountainous country, was to play it out with a haulage rope and winch it into place. The traction cable was usually installed in the same general manner, with the line resting on the tower sheaves. Once reaching full circuit, it was fitted around the driving sheaves, spliced together, and placed under tension. When all the cables were in place an empty car would be sent around the track. If all worked well, the remaining cars could be attached and put to work.[15]

Wherever possible, tramways ran by gravity, with just enough power being applied to set the line in motion. As long as loads went downhill, the system pretty much ran itself. In fact, many operators used the tramway's momentum to generate electrical power. At locations where gravity power was impractical, operators employed small electrical or gasoline motors to drive the machinery. Tramways also required communication between terminals. In the early days of tramming this may have been accomplished by hand signals, but bell systems (much like those used for hoists) later came into use. Eventually, telephone lines, either strung atop the tramway towers or on separate poles, became the most common form of communication.[16]

Once placed in operation, a tramway needed constant maintenance. A great deal of attention needed to be paid to the cables. During the first year after installation, cables stretched by about 1 percent. This required regular adjustments at tension points to prevent excess slack. Cable wear also created problems. Early operators discovered that a cable could wear out in as little as three months. And if wear was not detected, a cable might snap, dumping bucket loads of valuable ore on the ground. Single-rope tramways, which had permanently attached buckets, were particularly troublesome because the cable could not be rotated to assure even wear and thus they were prone to frequent failures. The Bleichert tramways, on the other hand, were designed so that operators could regularly turn the track cable to produce an even wear pattern. Trial and error taught mechanics to rotate the cable one-eighth of a turn with special wrenches every two weeks. The trick was to keep the cable in its new position despite a "tendency on the part of the cable to return to its former position." As a consequence, it was preferable to release tension on the cable before turning and then clamp it into place. In general, track cables experienced greatest wear in the vicinity of towers. Nevertheless, with proper turning and regular oiling track cables might last seven or more years. Although traction cables possessed a shorter life, they were relatively easy to replace. A reel containing the new traction cable could be set up behind the lower terminal. After releasing the tension, mechanics cut the old traction cable and attached a new cable to the end of the old one. The tramway was then run slowly, pulling the new cable into place. Meanwhile, a second reel took up the old line as it came down. When the new line completed its circuit, the two ends were coupled, placed around the main sheave, and put into operation.[17] Well-run and profitable operations tended to guard against wear and replace cables long before they presented a danger.

Accidents and mishaps occurred with some regularity and could be quite costly. In several instances, hasty construction and too much tension proved fatal, causing entire lines to collapse soon after entering service. This was the case at the Copper King Mining Company operation near Clifton, Arizona, whose owners spent heavily to build a two-mile tramway to connect its mine with a smelter. The tramway opened to great fanfare in the summer of 1891, operated for a few weeks, and fell down. A year later the company went out of business without ever repairing the line. Very much the same thing happened at the Mammoth-Collins gold mines near present San Manuel, Arizona, when twenty towers collapsed one morning. In this case, however, the line was rebuilt.[18]

Runaway buckets presented another danger. This was of particular concern on double-rope tramways. Numerous flaws might cause a car to jump the tracks, which could in turn pull others off the cable. If the traction cable snapped, the carriers essentially became free to roll downhill, crashing into each other until they fell off or piled up at the terminal. In one notable instance on the Blue Bell tramway in Arizona, the traction rope separated from a bucket, causing it to slide backwards down the line until it collided with the following bucket, which happened to be loaded with several cases of dynamite. Upon impact, both buckets fell to the ground, but fortunately no explosion occurred. In another potentially deadly accident, while working on the top of a tramway tower near Monte Cristo, Washington, mechanic Everett Borden suddenly looked up to see a loose bucket racing toward him. With no other way to keep from being hit, he jumped to the ground seventy-five feet below. Incredibly, a snowbank broke his fall and he suffered only a leg injury.[19]

Mother nature also hampered tramway operations. Despite claims to the contrary, bad weather, particularly snow- and windstorms, created numerous difficulties. Often located in severe high-altitude climates, the tramways were regularly showered with heavy snowstorms. Snow became more than just an inconvenience. Heavy snow often accumulated in buckets and froze the ore, especially if the line had to be stopped for any length of time. Occasionally snow drifted so high that buckets were knocked off the track cable. Avalanches also threatened operations, and it was not uncommon for the equipment to be taken out by snowslides several times a year. The tramways at Monte Cristo proved particularly vulnerable to winter storms. Almost every winter, avalanches knocked out one or more of the lines. In 1896, for instance, quick action by a repair crew enabled them to escape from a wall of snow when it smashed into the Mystery Hill tramway. Although the men escaped with their lives, the snow struck a cable with such fury that "the lower bull wheel, around which it ran, was smashed to pieces." Of course, not everyone lucked out. In early 1919 an avalanche swept down on Colorado's Sunnyside tramway, knocking out a number of towers and killing tram men Albert Wizorck and Lawrence H. Seay. Young Seay had only been on the job one day.[20]

The Sunnyside Mine lost one of its towers to snowslides so frequently that the company eventually built a fifty-foot tunnel into the mountain to store an emergency spare tower. When a snowslide carried one tower away, "the reserve tower is taken from the tunnel and erected at the site of the one carried away." The Camp Bird tramway near Ouray, Colorado, employed a special angle station to avoid snowslides. Although costly to construct, this unique station usually enabled the line to remain open during the winter "when several lines in the neighborhood suffered very serious damage and interruption of traffic." However, in 1907 a tremendous snowslide "started from the slope of Hayden's Peak above timber line . . . wrecking the supports under the tramway for about a half mile." Other means of contending with snowslides included the building of masonry snowbreaks, snowsheds, tunnels, and using extra-heavy timbers or steel supports.[21]

High winds also caused trouble. Suspended on long segments of cable, buckets tended to sway in the wind. To cope with swinging buckets in areas particularly susceptible to high winds, engineers built towers that separated the lines by ten feet or more. Yet even this adjustment could not deal with extraordinary situations. The Bleichert tramway at the Baker Mines in Oregon, for example, began to malfunction after a severe windstorm in March 1916. Upon examination, inspectors found a tangle of wire and buckets. At one point two buckets (one loaded, the other empty) had collided and locked together. At another location, the downhill track cable had blown across the parallel line, then attached itself to an empty bucket. Elsewhere, the traction cable whipped across an adjacent bucket and was resting on the grip lever. To repair this mess, mine superintendent Hamilton W. Baker rigged a platform to lower three men down the line. In a tricky and delicate operation that required twenty-seven hours, the tramway cables were finally straightened. After this work was completed, Baker remarked that the wind had been so strong that the "traction cable rubbed a tree 40 ft. to one side of the line of travel."[22]

Operating a tramway required skilled employees, and eventually a classification of workers called "Tramway Men" could be identified at many mines. On the larger systems, these men specialized in specific aspects of tramway operation. At Colorado's Silver Lake Mine, for example, in 1907 the company classified its tramway workers as grip men, bucket men, brakemen, oilers, and line men. The first three of these jobs paid $3.50 per ten-hour day, while oilers and line men took home fifty cents more. Comparatively speaking, these jobs paid about the same as underground or mill work. By World

War I, tramway operators at the Blue Bell Mine in Arizona (two per shift) earned $4.65 per eight-hour day.[23] It appears that most companies used similar wage scales, although not all mines assigned men full time to tramway operations.

Tramways were popular because they saved money over other forms of transportation. They required relatively few employees, generally moved in a straight line, and overcame natural obstacles. Actual operating costs, of course, varied according to the type of line, its capacity, length, the topography, and terminal needs. One 1917 study concluded that the cost of operation generally ran from one to five cents per ton-mile, with the cost being lower on longer lines because labor costs remained practically the same. "On first-class tramway equipments," the study continued, "the cost of repairs and maintenance is a very small item. For the first two or three years it is practically nothing, and after this period it would average from 3% to 5% of an amount equal to about two-thirds the value of the cable, machinery and timberwork. Power costs are at a minimum because in the great majority of cases a tramway will operate by gravity and develop surplus power." A more practical demonstration of potential savings could be seen when a group of mine owners in the Bald Mountain area of South Dakota proposed the construction of a seven-mile tramway to connect with the mills at Deadwood. At a total cost of $150,000, this line promised to cut in half what the owners were paying the railroads to move their ore.[24]

Many mining companies also used their tramways to haul supplies. This back traffic could be substantial and greatly improved the economics of tramway operation. Utilizing either empty returning buckets or special carriers, all sorts of materials went up as the ore came down. In his 1899 general report on the Silver Lake mines near Silverton, Colorado, Edward G. Stoiber detailed the great amount of back traffic going to his mines:

> The Upper Terminal, not only dispatches the Mine and Mill products, but receives goods for the Boardinghouse stores, and Mine and Mill supplies intended for the General Supplies Stores; Hay, Grain, Timber, Powder and Heavy Mine supplies, such as Rails and Lumber, may be unloaded on the Mine Main Level #1; also the coal intended for the Mill boiler coal bins; these arrangements provide for the economical handling of such material which runs up into the thousands of tons per annum.[25]

Yukon's Mountain Hero Mine brought timbers up from its wharf on Windy Arm by strapping them across a bucket. Since the line was designed to transport ore downhill by gravity, it became necessary to fill the buckets with a half-ton of waste rock if no ore was available to move the line. Once this ballast had provided the necessary momentum to move supplies (about 250 pounds per bucket) uphill, it was dumped near the lower terminal, creating huge mounds. The Mammoth Gold Mine in southern Arizona's desert carried an even more precious commodity—water. Located some three miles from the San Pedro River, the company tramway utilized returning buckets to haul river water to the mine and its thirsty workers. This operation lasted until miners struck water at the 800-foot level and flooded everything. Then they had too much water.[26]

The hauling of supplies also included foodstuffs for the men who worked and lived at the mines. Although most of the foods were ordinary staples, a company cook might occasionally bring in such treats as fresh fruit. In one amusing instance, a load of bananas intended for the bunkhouse kitchen broke open just as it was being loaded into a bucket. The tram crew naturally helped themselves to some of the delicacy before sending the leftovers upward. Halfway up, a tower repair crew spotted the bananas and reduced the load even more. Finally, men at the end of the line appropriated the remaining bananas, leaving the exasperated cook with nothing but a stalk. Contraband goods, especially whiskey, presented a more serious problem. Desiring to keep booze away from its isolated and bored bunkhouse residents, the Sunnyside employed a man to inspect the bedrolls of any worker riding the tram. The inspector accomplished his task by simply whacking each bedroll with a baseball bat. Nonetheless, some contraband got past the guard. In one case a miner, seeking to fool the operator, slipped a gallon of whiskey into the empty bucket going out ahead of him. This might have worked had not the operator at the angle station grabbed the bottle for himself.[27]

Although they were not intended for the purpose, aerial tramways also transported human cargo. From their earliest days, tramways held an irresistible fascination, although many companies discouraged the practice for safety reasons. Nevertheless, historical photographs show men riding in empty buckets or sitting on loads of ore. Aerial transportation proved an inviting alternative for workers who otherwise might have to walk or ride a mule to and from the mine. As T. A. Rickard noted in 1903: "In winter the managers of many of the properties find it expedient to make their trips to the mines over the tram route and in spring, when the deadly snowslide may launch itself down the mountain at any time, it is much safer to travel on the tramway, not because it is always immune from this peril, but because of the shorter time to which one is exposed to danger in making the journey on the tramway." The experience could also be thrilling. Among the very few firsthand accounts of riding a tramway is that of Martha Kelsey, reportedly the first woman to ride over Chilkoot Pass. Deathly afraid, she was sent out over the newly constructed line after being strapped into a freight box and reassured that all was safe:

> All at once, directly in front on me, loomed a great, black cliff, and I was
> dashed straight at it. I closed my eyes and shrieked as I never did before in
> my life, when, lo, the cliff was gone! I had been whirled just around its
> edge, and then I felt the awfulest sensation, for I was suspended over a
> great chasm, hundreds of feet above a glacial torrent, and it appeared to

A man riding a tramway bucket at Bingham Canyon, Utah, illustrates how precarious the ride could be. Note in particular how close his head is to the cable and the arm of the tower. Courtesy Utah State Historical Society, 979.2, p. 4.

be a mile from one side of the canyon to another, where the spiderlike cable lines were suspended from towers.[28]

In fact, by the late 1890s riding the tram was a rather common event. Andrew Hallidie experimented with carrying passengers on the 7,341-foot demonstration tramway he constructed on his farm in California's Portola Valley. A newspaper article in 1894 noted that in addition to material carriers, "there are three passenger carriers which will hold two passengers each and convey them above the ground across the chasms from the lower end of the Portola Valley to the top of the mountain, the distance being nearly a mile and a half, and the time occupied in transit is thirty minutes." Some mines were so isolated that miners lived at the bottom and were obliged to ride the tramway to work. Such was the case at the Shenandoah Dives Mine in Colorado. Many employees, especially the married men, lived in nearby Silverton and made a round-trip by tramway to work each day. One report indicated that some 250,000 passenger trips were logged annually on this line during

Construction workers gathered for a posed photograph on the Camp Bird tramway near Ouray, Colorado. Courtesy Colorado Historical Society, F7655.

the 1930s. In other cases, mining companies constructed bunkhouses near a remote mine entrance. Some of these structures were so precariously perched on mountainsides that men who tended to walk in their sleep were refused jobs. At these locations, tramways served as a major means of rotating miners in and out of their drab and dank living quarters. As boring as life in the camps could be, at least men residing at the top of the tram could walk to work.[29]

Others also had occasion to ride the high wire. During the violent Coeur d'Alene miners' strike in 1892, soldiers sent to quell the trouble rode empty buckets to and from the Bunker Hill and Sullivan Mine. Tramways were sometimes used to bring down injured miners and even the bodies of those killed in accidents, while the sheer thrill of riding a tram additionally attracted riders, especially young boys. It is unclear how widespread this practice might have been, but clearly some sympathetic employees let local lads ride the buckets, against company rules.[30]

Riding the tramway could be dangerous. The Smuggler-Union required riders to have a permit, which was good "only for employees while in actual performance of service for the . . . company." Nevertheless, personal-injury accidents occurred with some regularity. In her book *Tomboy Bride*, Harriet Fish Backus tells of a tram mishap that saddened the copper camp at Britannia Beach, British Columbia, while she was living there in 1910. A young company doctor, just arriving for duty and soon to be married, decided to take the tramway to look over the mine. Despite being warned not to raise up while passing the towers, he apparently lifted his head just as the bucket came upon a tower and was stuck in the head. Despite quick medical attention, he died later that day. Some thirty years later the owner of a Colorado mine riding in an empty bucket was knocked out and seriously injured when another bucket above him broke loose and slammed into his carrier. On another occasion, two miners departed the lower terminal of the Shenandoah Dives tramway for a ride to the mine. High above the Animas River Valley the machinery failed, leaving the men stranded as frigid weather set in. Rather than risk freezing to death, one of the men attempted to go hand over hand back down the cable, but lost his grip and fell to his death. The other man remained in the bucket and was eventually rescued. Another tragic event occurred on the nearby Sunnyside Mine tramway. The company followed a firm policy that no one could ride the tram if inebriated. One night four miners who had been drinking in Eureka were barred from riding back to the mine. Fearing they would be fired if they showed up late for work, they climbed a tower in order to jump into the buckets. One member of the group missed his bucket, grabbed the cable, and eventually fell into a canyon.[31]

In a more humorous situation regarding the Shenandoah Dives tramway, a prostitute from Silverton managed to jump in a bucket and head for the mine and its nearby bunkhouse, where many single miners awaited her arrival. Fearing a riot among his workers, the mine superintendent, who learned what was happening, arranged for the brakes to be applied as the lady neared the top, dumping her into a snowbank. She was picked up by mine officials, dusted off, and put on a return bucket, literally ridden out of town on a wire.[32]

In these ways, western mine tramways were built and operated. They hauled mine products ranging from gold, silver, and copper ore to gypsum and coal. Just about every major mining center could be counted upon to have at least one aerial tramway. Indeed, between 1890 and 1920 literally hundreds of these devices performed yeoman service. Among this group were some of the most spectacular pieces of machinery ever to grace the western landscape.

CHAPTER FOUR
GREAT WESTERN TRAMWAYS

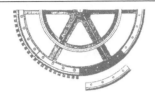

During the three decades between 1890 and 1920, aerial mine tramways reached a peak of popularity. Literally hundreds of these devices were installed, operated, and abandoned during these golden years. So common did they become that observers tended to take less notice, leaving future historians with little documentary information on the active life of many lines. Nonetheless, many of the most notable examples represented heroic feats of engineering or were so spectacular that they could hardly escape notice. Engineering journals, local newspapers, and manufacturers quickly noted anything unusual. While it would neither be possible nor useful to document every western tramway, this chapter will focus on the history of some of the more important operations, with emphasis on the Rocky Mountains, western Canada, and the desert Southwest.

THE ROCKY MOUNTAINS

Because of their rich mineral deposits, the Rockies attracted mining on an unprecedented scale. The rugged terrain and primitive conditions that accompanied high-altitude mining frequently necessitated the use of aerial transportation. Many notable systems were erected around the turn of the century, making the Rockies a center of tramway activity. Indeed, one could hardly visit any mining district in the region without seeing a number of tramways diligently moving ores and supplies back and forth. For the purpose of this study, the Rocky Mountain states of Colorado, Utah, and Wyoming will be featured.

The San Juan Mountains of southwestern Colorado, centering around Silverton but extending as far as Ouray, Telluride, and Rico, might be categorized as the tramway capital of the American West because of the high number of significant installations (at least two dozen). Although mining activities began in the 1870s, not until the arrival of narrow-gauge railroads during the following decade did the mines begin to prosper. By the 1890s hundreds of silver and gold mines were under development. Many of them required aerial tramways to transport ores from remote mine portals to the valley mills and ore bins that sprang up along the rail lines. During his famous tour of the San Juan region in 1903, T. A. Rickard commented on the proliferation of these tramways. They were, he suggested, an important feature of high-altitude mining, making it possible to tap virtually inaccessible loads. "We had already," he noted in his journal, "seen the tramways of the American Nettie, Bright Diamond, Grand View, Camp Bird, Smuggler-Union, Columbia, Liberal Bell mines, besides others, the names of which we did not know, so that with the group of three just referred to [Silver Lake, Iowa, and North Star], near Silverton, we had in the aggregate observed a good many examples of this kind of mountain engineering."[1]

One of the most active areas of tramway construction could be found along the twelve miles of road that follows the Animas River Valley from Silverton to Animas Forks. At least eighteen tramways operated along this stretch of valley, beginning with the Silver Lake basin just east of Silverton. This area had witnessed mining activity as early as 1872, when the Little Giant Mine constructed a small mill in Arrastra Gulch to process gold ore. Because of the rugged terrain, however, little headway could be made in tapping the rich gold and silver deposits as long as the veins could only be reached by mule train over dangerous mountain trails. Things changes dramatically in the 1890s. As Duane Smith notes, this decade "marked the coming of the end of an era dominated by graduates of the school of hard knocks. College-trained men moved to the forefront, to the dismay of many old-timers. They knew geology, mining engineering, and smelting." One of this new breed, Edward G. Stoiber, arrived in Silverton during the mid-1880s. Trained at the prestigious Freiberg School of Mines in Saxony, he first operated a custom sampler with his brother Gus. His attention, however, was quickly drawn to the Silver Lake lode. He immediately began purchasing claims and then set to work building one of the most modern mining facilities in the West. His initial efforts faced incredible obstacles, especially when he elected to build his mill complex next to the mine on the shore of isolated Silver Lake, some 12,250 feet above sea level. As one authority has noted, "His mine was above timberline, thousands of feet above the town, and accessible for years only by the steep mile-long trail which could only be

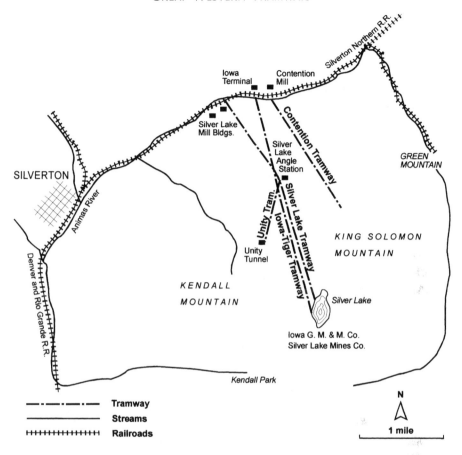

This 1914 map of the Silver Lake basin shows the location of the Silver Lake, Iowa-Tiger, Contention, and Unity tramways in the vicinity of Silverton, Colorado. Drawn by Amy Ashley.

traveled by mules. They dragged up the lumber and machinery that built the first concentrating mill at that high altitude."[2]

Although Stoiber's venture earned a profit, he recognized the need for a more effective process. In 1895 he contracted with the Trenton Iron Company for an 8,640-foot Bleichert tramway capable of moving five tons per hour. This line ran from the mill at Silver Lake down the mountain to Arrastra Gulch, from whence the concentrates could be hauled to a branch of the Silverton Railroad located along the Animas River. Once the tramway entered service it attracted notice for its long spans, one of which stretched almost 2,200 feet across a snowslide area. Shortly thereafter, Stoiber joined with J. H. Robin in organizing the Iowa Gold Mining and Milling Company to work some additional properties in the Silver Lake basin. Almost immediately,

This view of the Silver Lake Mine in the mountains above Silverton shows the Silver Lake tramway low on the hill. Higher up on the hill are the towers of the Iowa-Tiger tramway. Courtesy San Juan County Historical Society Collections.

they erected another Bleichert tramway that ran parallel to the Silver Lake line, ending in Arrastra Gulch where a mill was soon constructed.[3]

Despite the successful operation of these tramways, the mill at Silver Lake proved to be uneconomical. As a result, in 1898 Stoiber built a new mill on the east bank of the Animas River and abandoned the old one. This necessitated an extension of the tramway. Consequently, a second or lower section of 6,200 feet was connected to the original line by means of a combined midway terminal and angle station, which also accommodated a bend in the line. At the same time, the tramway's capacity was increased to thirty tons per hour. In an associated move, the company drove the Unity Tunnel some 600 feet under the Silver Lake basin and then connected the tunnel portal by tramway with the main line at the angle station. The adjacent Iowa tramway also added another section to its line after its owners consolidated with the Royal Tiger Mining Company. This lengthened tramway, usually known as the Iowa-Tiger, extended to a loading station on the Silverton Northern

Railroad, thereby creating the "longest straight tramway in the San Juans and one of the most spectacular." As a result, two major tramways, both owned by Stoiber interests, ran from the mines at Silver Lake to terminals along the Animas.[4]

These tramways were extremely efficient for their time. In describing his Silver Lake system in 1899, Stoiber noted its multipurpose advantages. Mine products were delivered to the railroad for transportation to the American Smelting and Refining Company's works in Durango, while every kind of supply went back up to the mine and mill. "The Lower Terminal, situated in the Freight Yards, with Railroad loading and unloading track on either side, is a four story building 50 feet by 100 feet ground space. In it is a Freight Elevator (run by Electric Motor) of a capacity of 1000 pounds, for raising supplies of all kinds from the lowest receiving floor, to the Bucket Dispatching floor." The same building contained eight large ore bins with a total capacity of 1,200 tons to store ores and concentrates should the railroad be closed by snow. The Silver Lake properties, with their tramways, proved very profitable, ranking among the greatest producers in the region and regarded as a sight not to be missed by visitors or investors. In 1901 Stoiber sold his properties for a handsome profit and the Silver Lake mines eventually came into the possession of the American Smelting and Refining Company. Ironically, the new owners never really turned a profit, although operations continued until 1914, when the mines shut down for good.[5]

Located a short distance up the river from the Silver Lake Mill one found the Contention Mill, built in the 1890s. This complex served the North Star and some smaller mines located near the top of Little Giant Peak. Identified as both the North Star and Contention tram, the 2.3 mile Desedau system descended some 3,200 feet to the riverside mill. It was notable because it connected at the top with a 2,000-foot, two-bucket (or jigback) tramway that delivered ore from the Big Giant Mine to the main line. This latter device worked by using the weight of the loaded bucket to lift the empty one.[6]

Ascending the Animas River road past Howardsville (and its tramway connecting the Little Nation Mine), one soon reached Eureka, site of the Sunnyside Mill. The Sunnyside Mine, located on the shores of Lake Emma some three miles to the northwest, had been discovered in the 1870s. Not until Judge John H. Terry acquired the property in the 1890s, however, did the mine become a profitable venture. Initially, Terry built a mill halfway between the mine and Eureka, the two points being connected by a tramway in 1897. The 8,600-foot line, designed by the Findlayson Company of Denver, was described as "a combination of both Bleichert and Huson, and this one has a capacity of 200 tons a day." Because the grades were too gentle, gravity could not power the line, thus requiring a small steam plant to supply the required momentum. In 1899 Terry opened a new mill at Eureka and

This view of the Silver Lake Mill along the Animas River shows the extensive mill, warehouses, and tramway terminal (just to the right of the twin stacks). Courtesy San Juan County Historical Society Collections.

extended the tramway some 7,100 feet at a cost of $15,000. For almost a decade, the Eureka Mill operated profitably, although it only recovered about 60 percent of the metal in the ore. In 1918 a new mill was constructed at Eureka, necessitating the installation of a new 16,000-foot tramway that ran from the mine to the mill, using several of the old towers.[7]

The new Sunnyside tramway, which operated until World War II, proved to be one of the most enduring installations in the San Juans. Utilizing more than a hundred buckets, it required an average of forty-five minutes to make the three-mile trip from mine to mill. The tramway also handled considerable back traffic, moving nearly all the supplies necessary to sustain the mine and the three hundred men it employed: "Special carriers are used for barrels and plates, . . . and sometimes loads are lashed between two buckets. Groceries, powder and almost all other forms of supplies are loaded in buckets." A lineman rode the system every day, making a general inspection and oiling the trolleys. A total of twenty-five men worked to keep the system in operation and had to stand ready to battle snow emergencies and other problems. The tramway also carried its own phone lines, which became particularly handy when storms wiped out other means of communication. Famous for its battles against snow, the Sunnyside's tram often proved the only connection with the outside world.[8]

Another four miles on up the Animas Valley came the boom camp of Animas Forks, where in 1906 the Gold Prince Company constructed the largest and most modern concentrating mill in Colorado. In addition to other local mines, the mill served the Gold Prince Mine located at the head of Mastodon Gulch at an elevation of 12,500 feet. The mill presented an imposing sight. As one observer wrote: "It loomed up against the landscape as huge as a small mountain. The steady pounding of the hundred stamps made a din and rumble like distant thunder as it kept up its ceaseless work of pounding out the silver and gold." The Gold Prince, and its equally impressive tramway, would prove to be one of the most expensive operations in the San Juans, and one of the most disappointing.[9]

Cyrus Davis and Henry Soule, owners of the Gold Prince, had long been fascinated with the area's potential. They had purchased the old Sunnyside Extension (which once used a Huson tramway), renamed it the Gold Prince, and in 1904 set about developing it. To fund the mill and develop the mine, the owners attracted investors with a promise of steady profits. One stockholder's report estimated that the mine contained enough precious metals to keep the Gold Prince in operation for sixty years. Thus, in 1906 the company paid the Denver Engineering Works $75,000 to install a 2.37-mile Bleichert tramway connecting the mine and mill. Engineer W. Z. Kinney supervised the construction.[10]

The Gold Prince tramway was worthy of the attention it attracted. It ran from the mine entrance down and diagonally across Mastodon Gulch, then climbed to the crest of Treasure Mountain to an angle station. At this point the buckets made a thirty-degree turn by means of a short curved rail section, which then connected with a second track cable taking the load the rest of the way to the mill. The thirty-three tramway towers and angle station were constructed of heavy timber supported on concrete footings. Even today, nearly a century later, the surviving towers are impressive. At the huge lower terminal, buckets capable of carrying 1,500 pounds each and traveling at a rate of 250 feet per minute automatically dumped their load into large steel bins. In all, the tramway was capable of delivering sixty tons of ore per hour. The entire system was powered by a forty-horsepower electric motor. It also seems probable that power lines stretched across the towers, carrying electricity from a power plant at Animas Forks to the mine.[11]

Ironically, the Gold Prince never proved very successful. How long the mine operated is uncertain, but it closed before 1917. About that time the Sunnyside Mining Company purchased the property. They dismantled the

This 1999 view shows all that remains of the Gold Prince tramway and mine. The mine was located in the center of the picture where the two small snow-filled gullies come together to form a V. The Sound Democrat Mill and tramway terminal can be seen at the left. Courtesy Glen Crandall Collection.

tramway, taking the cable and machinery to Eureka where it was apparently used as part of the second Sunnyside tramway, which was installed in 1917.[12]

Smaller tramways also operated in the same area. For the most part unremarkable systems of simple construction, only one deserves a few comments, primarily because it connected with what is the last standing stamp mill in the district. The Sound Democrat Mine and mill, located near the Gold Prince Mine in Mastodon Gulch, operated a 1,100-foot Bleichert tramway. Being of small capacity, it employed a track rope one inch in diameter and a half-inch traction cable. There may have been only one support tower and it probably utilized just a few buckets. Considering that the mill only housed seven stamps and never produced much, it obviously did not need great capacity. Nevertheless, with the mine portal located some seven hundred feet above the mill, the tramway proved absolutely essential. Built in 1905–1906, the Sound Democrat operated for only a few years.[13]

Located northwest of the mines of the Animas drainage and over the mountains near the town of Ouray, one found another pocket of mines using aerial transportation. Most notable was the fabulously rich Camp Bird Mine. Owned by Thomas Walsh, who spent his money lavishly, the Camp Bird became one of the most modern and profitable gold operations in the San Juan Mountains. Sometime around 1896, Walsh installed a tramway that T. A. Rickard later described as "one of the best installations in Colorado." The Camp Bird tramway, a Bleichert system manufactured by the Trenton Iron Company, extended some 8,550 feet, with an angle station and a fall of 1,840 feet. Considered to be "a thorough piece of engineering work," each of its forty-six buckets carried 750 pounds of ore. As the Trenton Company boasted: "The cost of transporting ore, in this case, with an output of 3,100 tons per month, is fifteen cents per ton, but this includes the carriage to the mines of supplies of all kinds which are loaded into empty returning buckets. This feature of mountain tramways is most important because during several months the trails are often impassable by reason of snowslides." This tramway cost $55,094 to install.[14]

Rickard also commented on the Leschen tramway installed at the American Nettie Mine, located high on a ledge northeast of Ouray. Less than a mile in length, it was nonetheless "a picturesque bit of engineering," crossing a canyon with a 2,100-foot unsupported span. Because of the very abrupt contour of the land, the tramway line descended from the loading terminal at a very steep angle that required special safety measures. Rather than using friction grips, the line employed permanently attached clips to hold its two buckets. It also sported a mechanism to "diminish the vibrations attendant on the removal of the load from the line and the return of it into service." The American Nettie was frequently touted as the highest in the world.[15]

Just a few air miles from Ouray, although a great distance by railroad, were Telluride and the great mines near Pandora. Here one found the rich Smuggler-Union and the famous Tomboy. These two mines presented a stark contrast in regard to mill location. Owners of the Tomboy constructed their mill near the mine, thus eliminating the need for a tramway to move ore. The Smuggler-Union, on the other hand, elected to construct its mill at Pandora and send the ore down by tramway. Because the Smuggler-Union worked a rich vein nearly a mile long, engineers drove the Bullion Tunnel through the mountain. The double-rope tramway ran from the portal of the tunnel to the mill. Eventually, the Tomboy also built a tramway so that mill concentrates could be transported to loading bins on the Rio Grande Southern Railroad branch at Pandora. Although these tramways were efficient, the area was prone to devastating snowslides. In the spring of 1902, for example, the Smuggler-Union tramway was put out of action for several weeks by an avalanche. As T. A. Rickard observed: "The destructiveness of a snowslide must be seen to be appreciated; buildings and tramways are as toys before its fierce oncoming and men in the path of its descent are as straws in a whirlwind."[16]

Other nearby areas in the San Juan Mountains also required major tramway systems. In the Gladstone district (north of Silverton) the Gold King, Mogul, and Henrietta mines all operated substantial tramways. In Cunningham Gulch, east of Howardsville, tramways were operated by the Old Hundred, Buffalo Boy, and Green Mountain mines. The small town of Rico was known for years as the location of a tramway that ran right down the street. Further to the southeast at Creede, the rough mountain terrain mandated that the Smuggler, Bachelor Commodore, and Amethyst, as well as several smaller mines, used the wire devices. The Kentucky Bell Mine at Creede sported the world's longest "free swinging" jig-back tramway of its day. It stretched directly over Willow Creek Canyon and the town of Upper Creede.[17] Additional tramways could be found at other locations in the Colorado Rockies; indeed, too many to document.

Aerial tramways were not nearly as numerous in Utah, although several notable examples were built. Tramways became necessary as Utah miners began to develop the rich deposits of copper and coal located in the Wasatch Mountains and just south of the Salt Lake basin. Particularly interesting were the tramways serving the copper mines in and around Bingham Canyon. Located just thirty miles southwest of Salt Lake City, the area had been mined for gold and silver since the 1860s. Not until the end of the century, however, did interest shift to copper, which soon emerged as the major product. Between 1898 and 1910 the Trenton Iron Company built four different tramways to move copper ores from the various mines to local smelters. As would be amply demonstrated with these applications, in mountainous country tramways would prove more economical than surface rail lines, which "could only be built along a circuitous route, involving great expense for grading and trestlework."[18]

The first tramway was constructed in 1898 to connect the Highland Boy's No. 7 Tunnel with receiving bins located along the Denver & Rio Grande Railroad (from whence ore was hauled to a smelter in Garfield). With a straight length of 12,700 feet and a fall of 1,120 feet, this tramway operated by gravity, supplemented only by a small engine for providing power to run the line without loads. Initially designed to deliver twenty-five tons per hour, the continued development of the Highland Boy Mine caused the capacity to be increased to forty tons, which was accomplished by speeding up the line. This system also featured some unusual tower construction. At one point, where a rock face could not be blasted away (because of some houses below), a tower support had to be balanced on the rock itself. The cost of operating the Highland Boy line over a span of eight years, including repairs and replacement cables, came to less than seven cents per ton. This tramway operated until 1910 when a new one took its place.[19]

In 1903 the United States Mining Company opened a 11,450-foot tramway to connect the Evans Tunnel (which served the Telegraph Mine) with ore bins situated along the railroad. At the same time, a 4,025-foot branch tramway line extended from the Evans Tunnel terminal to the Old Jordan Mine. This installation proved to be quite complicated. The rugged topography required several sharp vertical bends served by rail stations. Another rail section became necessary to carry the line above a cave-in. To prevent the line from being closed by snow, all the rail stations were covered. Since the line crossed several summits before descending to the terminal, a uniform speed was maintained with several fifty-horsepower hydraulic controllers. It was particularly gratifying to the Trenton Company that the tramway successfully competed with a railroad line that ran within two hundred feet of the Evans Tunnel. As William Hewitt remarked: "The satisfactory service rendered by this line furnishes a marked instance of the economy of aerial tramways over surface railways in mountainous localities."[20]

Much the same could be said for the tramway installed in 1907 to connect with the Yampa Smelter. Paralleling the Highland Boy line, this system was not authorized until after a rail line had been built, only to prove too expensive to operate. Only then did the owners consider a tramway, which quickly reduced costs to a reasonable figure. Although the system required some thirty-two towers and four tension stations, "the reduction of the cost of handling the ore by aerial tramway instead of by railroad has been a very marked one in this case." One particular economy involved eliminating the loading and unloading of railroad cars by hand labor, thereby saving about thirty-five cents per ton.[21]

Without question the Bingham Canyon tramway that attracted the most attention was the 1910 line that replaced the original Highland Boy tramway. This mammoth system ran from the Utah Consolidated Mining Company's Highland Boy Mine to the new International Smelting and Refining Company plant at Tooele, a distance of over four miles. Because of its great length the line was divided into three sections. Altogether the system required eighty substantial wooden towers set on concrete piers, two control stations, six tension stations, and one rail station. Since the tramway was built for high-speed operation (600 feet per minute) and held 1,500-pound-capacity buckets, everything needed to be especially sturdy. Tower caps (which supported the cables) were constructed "of 10" x 12" timber substantially supported by posts and side braces." Each control station housed a 100-horsepower General Electric induction motor for starting the line in motion. Once moving, gravity took over, enabling the motors to generate power: "When the tramway is loaded and in operation the motor at No. 2 control will act as a generator, No. 1 being cut out, and will develop 100 h.p. which will be used at the mine." These motors could also be used as a brake on the system. Overall, this massive tramway, which delivered 100 tons of copper ore per hour to the

One of the Trenton tramways at Bingham Canyon, Utah, stretches over town on its way from mine to smelter. Courtesy Utah State Historical Society, 979.2, p. 49.

Tooele smelter, proved to be extremely efficient. As Leroy A. Palmer predicted for *Mines and Minerals* magazine, the tramway would "effect a savings in freight of 75 cents per ton or $600 a day [versus rail transportation], practically $220,000 per year, and thus return the cost of construction within a very short time."[22]

Utah coal mines also used aerial tramways. A particularly unique system was constructed by the Spring Cañon Coal Company, located at Storrs, about five miles from Helper. Although only 3,000 feet in length, the rugged topography of the area caused designers to display considerable ingenuity. The mine itself was located on a rocky ledge and the tramway carried coal to a rail siding some 321 feet below. At one point, "where a sharp change of grade takes place, the vertical curve in the cable is divided between four saddles. One side of this structure rests on solid rock, while the other is cantilevered over the edge of the precipice." The most unusual feature of the tramway, however, were the buckets. They were designed so that two of them at a time could be placed on a set of wheels (trucks) and be used as regular mine cars, each with a capacity of about two thousand pounds. In this way they were loaded inside the mine and brought to the loading terminal by electric locomotives. Once at the terminal, the cars were pushed onto a special table that

Tramway towers and smelter on the sixteen-mile system erected at Grand Encampment, Wyoming, in 1904 by Leschen & Sons. Note that towers were not anchored in concrete. Courtesy Wyoming Division of Cultural Resources, Stimson 762.

raised the car while a pair of hangers hooked onto the bucket. As the table lowered again the wheel set came off and the bucket was sent down the line. The only drawback to this ingenious system was its limited capacity. Mine owners desired an hourly rate of 250 tons, more than double the original rating. Eventually the manufacturer agreed to upgrade the system to handle the increased tonnage.[23]

Located at Grand Encampment in southern Wyoming near the Colorado border was the longest tramway in the United States. Some sixteen miles in length, it was constructed for the North American Copper Company by A. Leschen & Sons in 1904. This system proved unusual in a number of ways. In the first place, North American Copper wanted the line designed and placed in operation within seven months at a location that presented some unique construction challenges. With neither end of the tramway located near a railroad, all materials except timber had to be hauled in by teams. Leschen accepted the challenge and assigned Byron Riblet, who was then

Tramway transfer station No. 2 (where two sections joined) on the sixteen-mile system erected in Carbon County, Wyoming, at Grand Encampment. Courtesy Wyoming Division of Cultural Resources, Stimson 751.

in their employ (see Chapter 2), to design and build the system. Riblet quickly surveyed the 84,500-foot route and immediately began shipping in materials, meanwhile cutting the required timber in a nearby forest. To expedite construction, Riblet then hired a hundred Canadian carpenters and set up several camps along the route to put up different sections simultaneously. In a little less that seven months the Grand Encampment tramway entered service, carrying 400 tons of ore per ten-hour shift to the smelter.[24]

The Grand Encampment tramway represented a mechanical marvel. To extend over sixteen miles, it became necessary for engineers to build the line "in four separate and distinct sections, each being independently operated by separate engines." Tension stations were placed at one-mile intervals to keep the line taught. A total of 304 towers were necessary to support the cables, their placement intervals ranging from a few feet to 2,300 feet apart. Each bucket held 6.5 cubic feet of ore and was set up for automatic operation, thereby eliminating the need for more than a couple of men to oversee the entire operation. As one contemporary article noted: "The principal advantages claimed for this system are a perfect clip, automatic operation, low cost of maintenance and operation, absolute safety, and large capacity." Unfortunately, the Grand Encampment tramway never proved particularly successful. Early on, a problem developed with the way the buckets were attached to the traction rope: "the track ropes were so slack that the buckets in passing over the long spans would hang down so low" that the buckets could break loose and run away. When this occurred, the lug attached to the traction cable began whirling and would snag on the next tower: "towers were all set on the ground without any anchorages, and consequently it dragged this tower to the next one and then hooked onto that one and pulled it to another one." Because the operator had no indication of the problem before the line stopped, "as many as four towers [became] all piled up in one bunch." Then in 1906 a fire destroyed the smelter's concentrating mill and the rest of the plant came down a few years later. Nevertheless, the tramway was operated for several more years, delivering ore in case the mill reopened. It never did, and in 1910 the tramway was dismantled.[25]

THE DESERT SOUTHWEST

Aerial tramways were less visible among the mining operations of the Southwest. Although mining in desert regions may appear to have had less need for tramways, many locations were in fact quite rugged. With far fewer tramways than in the Rocky Mountains, the arid land around Death Valley, California, and in central Arizona nevertheless had a fair share of interesting systems. Operating in an environment where snow and altitude presented few problems, the desert offered its own set of unique challenges.

Surprisingly, perhaps, Death Valley in eastern California hosted a number of mines that required forms of aerial transportation. A recent National Park Service inventory of low-flying aircraft hazards (cables) identified over twenty different tramways or cableways.[26] Many were small lines running only a few hundred feet, but several more spectacular efforts did exist in the area.

One of the most interesting Death Valley tramways was operated by the Keane Wonder Gold Mining Company. This operation proved to be one of the most successful gold producers in the region. Initially discovered in 1904, it could not be fully developed until 1907, following several changes in ownership. After building a twenty-stamp mill at the base of the Funeral Mountains, the owners quickly turned their attention to the construction of a double-rope tramway to carry ore down from the mine, about a mile distant. Despite such intense summer heat that workers could only work during the mornings and evenings, an elaborate upper and lower terminal and some thirteen towers using nearly 75,000 board feet of lumber were constructed in only four months. The tramway entered service in October 1907. Known locally as the "sky railroad," the gravity-powered system operated almost continually for five years before the gold played out. Several later owners attempted to rework the mine and as late as 1940 the rebuilt tramway was still operable. By 1942, however, all mining had ceased and the mill was taken down. Fortunately, the Keane Wonder tramway escaped destruction and remains today a rustic monument to a remarkable era in mine technology.[27]

Despite its desolate location, the Keane Wonder tramway resembled many other western tramways. Such could not be said about the Saline Valley salt tram, which was regarded as "one of the most novel aerial tramways ever constructed." This tramway formed an important part of one of the most unusual mining ventures in Death Valley. The Saline Valley, as the name implies, contained vast fields of halide (common table salt). In 1911 the Saline Valley Salt Company began to seriously develop the salt deposits. To move the product to the nearest railroad connection, the company—after ruling out a pipeline or railroad—contracted with the Trenton Iron Company to construct a 13.5-mile double-rope tramway across the Inyo Mountains and down into the Owens Valley to meet the Southern Pacific narrow-gauge railroad at "Tramway," about four miles north of Keeler. Moving salt from the flat lake bed over the 8,740-foot summit of "one of the highest, steepest, and roughest ranges in the desert," required considerable engineering. In all, the lengthy system used five intermediate control stations, twenty-one rail stations, and twelve anchorage-tension structures. Construction began in September 1911 and immediately ran into difficulties. Some of the terrain around lower Daisy Canyon was so rugged that contractors needed to erect a temporary tramway to move materials and water to the construction site. In all, this remarkable system required over a million board feet of timber and fifty-four miles of cable. It finally entered service on July 2, 1913, to the cheers of local dignitaries. Working conditions, however, were brutal. No bunkhouses were built at the salt fields. Workers, mostly local Indians and Mexicans, apparently lived in tents and worked in temperatures of over a hundred degrees. Perhaps the best job on the line was that of the control operator stationed at the line's 8,720-foot summit where it was cool in the summer.[28]

The half-million-dollar salt tram employed 286 specially designed buckets fitted with covers to keep moisture out. Because of a failure to take into account that the salt cargo might contain moisture and thus weigh more, the traction cable grips slipped on the excessively steep grades if the buckets were fully loaded, a major concern since profit margins on the salt depended on delivering the largest possible volume. Unfortunately for the company, it required two years to correct the grip defect, during which time there were numerous accidents and rope failures. The necessity of operating at two-thirds capacity took such a financial toll on the company that in 1916 it leased the operations to the Owens Valley Salt Company. Although the tramway operated at full capacity for the next four years, delivering between twenty and thirty tons of salt daily to the railroad, profits remained elusive (in 1917

Taken about 1920, this photograph shows some of the promoters on the Salt Tram hamming it up for the camera. From this loading terminal in the Saline Valley, the buckets traveled some thirteen miles, rising from an elevation of 1,050 feet to crest the summit at an elevation of 8,720 feet before descending to the discharge terminal at Tramway (Swansea). Courtesy Eastern California Museum, HES 2.

a 7.5-pound box of high-grade table salt sold for ten cents). In 1920 the Trenton Iron Company repossessed the system and in 1928 sold or leased it to the Sierra Salt Company. After a complete renovation, the system began to ship up to 100 tons per day. Even this did not work. Salt prices collapsed with the onset of the Depression and Saline Valley salt mining ended. Described as "one of the two or three most monumental structures ever built in the Death Valley region," several of the towers and at least one of the control stations remain standing.[29]

Located just four miles south of the salt tram's terminus along the Southern Pacific narrow-gauge is the town of Keeler, terminal site of another important tramway—the Cerro Gordo. The silver-lead discoveries at Cerro Gordo, some eight miles up in the Inyo Mountains, dated back to the 1870s. In the early days of production, ores from the mine were carried across Owens Lake by steamboat and then hauled by mule team to Los Angeles, where some of the high grades were then forwarded to Wales for processing. After several stops and starts the mine witnessed a brief comeback in 1908. At this time a tramway was constructed between the mine and ore bins located at Keeler, which was now served by the railroad. Soon failing, the operation was reopened in 1915 to mine zinc. During the summer of that year twenty-five

The Death Valley tramway's 286 buckets were specially constructed with a cover to prevent moisture from getting to the salt. This scene shows the Owens Valley terminal. Courtesy Eastern California Museum, WIL 103.

men reconditioned the 29,100-foot tramway and ore bins, and built a new control station at the halfway point. By September the tramway was delivering 100 tons of ore per day to Keeler at a savings of $2.50 per ton over earlier transportation. The line's steep grades permitted it to be powered by gravity, the most difficult problem being to brake the buckets as they sped downhill. Zinc mining at Cerro Gordo lasted until the Depression. In 1959 the tramway was dismantled and moved to nearby Candeleria, Nevada.[30]

A number of tramways operated in Arizona. Although the Copper State could claim more than two dozen aerial tramways, only a few attracted much attention. Two of the most interesting tramways were located in the Bradshaw Mountains north of Phoenix. Similar in construction, both the De Soto and Blue Bell tramways were connected to the Bradshaw Mountain Railway and owned by the Consolidated Arizona Smelting Company. They appeared at the height of the Arizona copper boom and managed to survive until the 1920s.

The De Soto Mine tramway went up first. Located on a high plateau, the mine site had been known to contain copper since the 1870s. A lack of

This photo shows the Keeler, California, terminal of the Cerro Gordo Mine tramway just before it was dismantled in the late 1950s. Courtesy Glen Crandall Collection.

access, however, had made it impossible to successfully develop the property, owing to the fact that ore had to be packed all the way to Prescott. The completion of the Bradshaw Mountain Railway to Middleton in 1903 put the mine less that a mile from the railroad, offering an opportunity to develop the De Soto. This was especially convenient, since the railroad, as it wound its way toward Prescott, passed through Humboldt, where Consolidated Arizona operated a smelter. This led Consolidated Arizona to purchase the De Soto in 1904, refurbish the mine, and construct a .75-mile aerial tramway to connect the mine with a siding at Middleton. Completed in April 1904 using machinery and cable supplied by the Trenton Iron Company, the double-rope tram would deliver about three hundred tons of ore per day to the Middleton loading bins for shipment to Humboldt.[31]

A year later, Consolidated Arizona acquired the even more productive Blue Bell Mine, located about five miles north of the De Soto. Quickly upgrading the property, the company installed a 3.3-mile Trenton tramway, running in a northerly direction to connect with the Bradshaw Mountain Railway at Blue Bell Siding, a few miles from Mayer. Thus, although the two

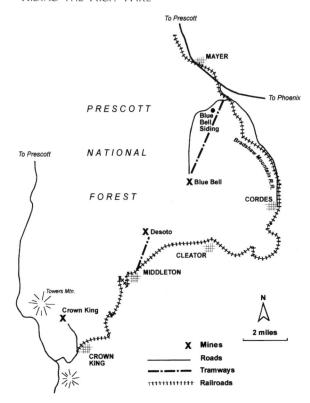

Location of the Blue Bell and De Soto tramways in Yavapai County, Arizona, both owned by the Consolidate Arizona Smelting Company. Each line went in a different direction to connect with the Bradshaw Mountain Railroad. Drawn by Amy Ashley.

mines were only a few miles apart, their tramways went in opposite directions to connect with the same railroad. At Blue Bell Siding the Consolidated Arizona constructed a large wooden tramway terminal. The Blue Bell tramway proved to be a good investment. At a capacity of 500 tons per day, it delivered ore to the railroad at a cost of 5.6 cents per ton mile. This figure included wages for two operators per shift, at a daily rate of $4.65.[32]

Typical of most Arizona tramways, the De Soto and Blue Bell systems operated on and off for several years, their use dependent on the vagaries of the copper market and the quality of ore reserves. The De Soto line operated from April 1904 to August 1907, and again from 1915 until the smelter at Humboldt shut down in 1920. Beginning in 1922, Consolidated Arizona's successor, Southwest Metals, subleased the mine to a succession of small operators who may have used the tramway for short periods before a fire in 1930 destroyed the Middleton terminal. A year later the mine closed for good. The history of the Blue Bell line followed a similar chronology. After closing in 1907, it reopened in 1913 with a completely rebuilt tramway line, which ran until 1921. It apparently operated again between 1924 and 1926, by which time the mine had played out. Although Southwest Metals subleased the

84

property, the tramway had deteriorated to such a degree that it became inoperable. In was scrapped in 1932.[33]

A fairly typical example of early-century Arizona tramways was located at Christmas in Gila County. Although copper claims in the area dated back to 1879, little mining was done until after 1900 when the Saddle Mountain Mining Company began operations in the rugged desert region. While waiting for a rail connection to arrive from Winkleman, in 1907 the company built a 3,600-foot aerial tramway from the mine at Christmas Camp to bins located on the banks of the Gila River. This construction unfortunately coincided with a slump in copper prices, which idled the mine and eventually led to its purchase in 1909 by the Gila Copper Sulphide Company. In 1911 the Arizona Eastern Railroad completed a line up the river to a terminus at the mouth of Copper Canyon (about a mile and a half from the mine). Once again, however, financial problems shut the mine down. It reopened in February 1916 under lease to the American Smelting and Refining Company. To work the mine more effectively, in April 1916 the company replaced the old tramway with a 7,300-foot Bleichert system that connected the Christmas Mine directly to the Arizona Eastern Railroad terminal. The tramway itself dropped some 950 feet at a rated capacity of fifty tons per hour. It also turned an electrical generator, which provided power to the mine. The upper ore bin held 250 tons, the bottom one 1,000 tons. Cost of operation in 1919 was reported to be 17.5 cents per ton. Ore delivered to the terminal was hauled by rail to the ASARCO smelter at Hayden. Like other Arizona copper mines, the Christmas operated sporadically after 1920, although it did not close entirely until 1977. Exactly what happened to the tramway is not clear. It was most likely used during the 1920s to ship ore to the railroad and probably became inoperable by 1930. By the time Inspiration Copper Company converted the property to an open pit in 1966, all remaining traces had vanished.[34]

WESTERN CANADA

The silver mines of the West Kootenay region of southeastern British Columbia represent another area of significant tramway construction. This was particularly true for the Slocan Valley, encompassing the Sandon and Silverton mining districts. These towns became mining hot spots during the 1890s and remained active for several decades. Eventually over 150 silver mines were developed. The first tramway in West Kootenay was the 1895 Hallidie system built at the Silver King Mine near Nelson (see Chapter 1). Several other single-rope tramways also existed in the region, including a small system at the LeRoi Gold Mine near Rossland. Nonetheless, once the silver market picked up around the turn of the century, mine owners turned to the more modern double-rope systems.[35]

The Riblet Tramway Company dominated the tramway business in West Kootenay, reportedly building between fifty and sixty aerial lines in the

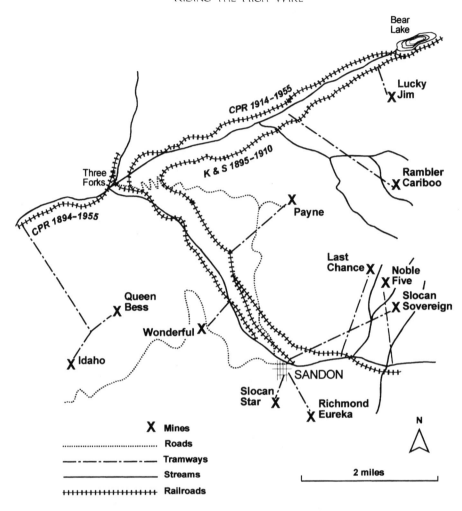

Bear
Lake

CPR 1914-1955

Lucky
X Jim

Three
Forks

K & S 1895-1910

Rambler
X Cariboo

CPR 1894-1955

Payne

Last
Chance X

Noble
X Five

Queen
X Bess

Slocan
X Sovereign

Wonderful X

X Idaho

SANDON

Slocan
Star X

Richmond
X Eureka

N

X Mines

.... Roads

— · — · — Tramways

——————— Streams

++++++++++++++ Railroads

2 miles

This map illustrates the proliferation of tramways surrounding Sandon, British Columbia. Drawn by Amy Ashley, based on a map in Dave May, Sandon: The Mining Centre of the Silvery Slocan, *1986.*

region between 1896 and 1920. Although the company was headquartered in Spokane, Washington, it maintained an office and warehouse in Nelson and a facility at Sandon. These branches manufactured various replacement parts and repaired local tramways. After building his first tramway at Sandon's Noble Five Mine in 1896–1897, Byron Riblet put up a number of additional tramways in West Kootenay before temporarily going to work for A. Leschen & Sons. By 1908, however, he returned to Spokane to run his own company, again focusing heavily on West Kootenay.[36]

The area surrounding Sandon proved to be incredibly rich, with most of the mines being located in the mountains high above the town. Before the arrival of railroad lines and tramways, the transportation of large amounts of ore by horse or mule team proved very expensive. The trails were also notoriously unreliable. Once snow began falling, the trails become unusable until the snow could be packed down, thereby causing the mines to close down. Much the same thing happened in the late spring as melting snow turned the paths into quagmires. Tramways solved this problem to such a degree that most of the major mines soon opted to install the necessary machinery. Eventually a dozen tramways were built in the general vicinity of Sandon. Almost all of them operated by gravity and many were spectacular. Towers often approached forty feet in height and one tower on the Slocan Sovereign tramway reached almost a hundred feet into the air. One cable line stretched unsupported across a span of 2,800 feet. The Sandon tramways had various capacities, with most of them being equipped with Riblet's patented automatic loader.[37]

Not only did these tramways carry ore down the steep mountainsides, they took supplies up to the mines. And despite Canadian regulations to the contrary, riding tramways appears to have been a common practice. Several women and children, "apparently much braver than [their] male companions," were among the first passengers. Nonetheless, the practice could be dangerous. If the traction rope jumped a sheave a passenger might be stranded high above the ground. While most riders would wait for repairs, at least one man on a Sandon tramway, caught hanging 175 feet in the air, "panicked and slid down a rope to the ground, burning his hands so badly that he couldn't work for a long time." On occasion the tramways also had the somber task of carrying down the bodies of men killed in mine accidents.[38]

At nearby Silverton, on the shore of Slocan Lake, the Standard Mine and Mill also turned to the Riblet Company for tramway construction. Described as "the most illustrious mine in the entire Slocan country," this operation produced more silver (9 million ounces) than any other mine in West Kootenay. Sometime about 1911, the company built a large concentrator at Silverton to process the rich ore. At the same time, it ordered a 7,900-foot tramway to be erected between the mine's No. 6 level and the mill. Employing standard Riblet technology, the tramway relied on wooden towers and carried timbers and other supplies to the mine on special carriers. Several other mines in the area also used Riblet tramways. The Van Roi Mine, which produced silver, lead, and zinc, operated a 4,900-foot tramway, while the adjacent Hewitt-Lorna Doone Mine utilized two lines. Including other tramways in the vicinity, the Silverton district may have been second only to Sandon in the number of tramways.[39]

Another area where the Riblet Company built a number of tramways involved the short-lived Windy Arm silver boom of 1905–1906. Developed

Conrad mines tramway terminal on the shore of Windy Arm, Yukon Territory, in 1906. This lakefront terminal served the Mountain Hero Mine. Courtesy J. B. Tyrrell Collection, Thomas Fisher Rare Book Library, University of Toronto.

primarily by promoter John Henry Conrad on the southern border of the Yukon Territory about halfway between Skagway and White Horse, the mines of Montana Mountain first appeared to be incredibly rich. In a spirit of high optimism and in search of investors, Conrad quickly ordered a tramway from Royal N. Riblet for his Mountain Hero Mine. In July 1905 Conrad paid Riblet $80,000 to erect an 18,697-foot double-rope system running from a wharf terminal on the shores of Windy Arm to the mine, a climb of 3,464 feet. Riblet assigned R. E. Lanyon, of Nelson, British Columbia, to personally supervise construction. All the necessary timbers and machinery were on hand by August, along with a fifty-man Canadian construction crew brought in by Lanyon. The Mountain Hero tramway presented a number of engineering challenges. In particular, the rough terrain required several long spans between towers. One in fact, some 2,960 feet in length, would be the longest unsupported cable span in the world at the time. As it turned out, construction required more time than anticipated, despite a personal visit from Riblet

Montana Mine tramway loading terminal, Yukon Territory, 1906. Note timbers arriving on an ore bucket, one of the many adaptions made to carry supplies. Courtesy J. B. Tyrrell Collection, Thomas Fisher Rare Book Library, University of Toronto.

in September. The following month, while attempting to string the 90,000 feet of steel cable, workers broke a capstan, which could not be immediately replaced, causing Lanyon and his crew to return home. Indeed, much to Riblet's chagrin, so many problems surfaced that the tramway did not become operational until June 1906. By this time prospects for the mine had diminished and the tramway actually did more work delivering supplies to the mine than it did in moving ore.[40]

Well before the Mountain Hero tramway entered service, several other area systems were under construction. By August 1906 a 1,850-foot tramway connected the Venus No. 2 tunnel with a terminal on the lake. Soon thereafter, a concentrator was built at the terminal site. Meanwhile, another tramway connected the Vault Mine with the shore. Unfortunately, the Windy Arm stampede quickly petered out, and by 1908 mining was all but dead. John Conrad lost interest, investors lost money, miners moved out, and the tramways were either abandoned or sold off.[41] As elsewhere in the West, the life of a tramway depended on the quality of ore deposits and the price of metals. That many tramways, so carefully constructed at heavy expense, should so quickly be abandoned is a testimonial to the dreams of mining men like Conrad who always thought in grandiose terms.

There were, of course, other significant aerial tramways in the mining West. Montana, Washington, the Black Hills, Idaho, Oregon, New Mexico, northern California, and Nevada all had examples of interesting and successful cable technology. Collectively, the aerial tramways utilized at western mine sites saved operators millions of dollars and helped move billions of tons of mineral product to market while lifting equally impressive amounts of supplies and materials to the mines. Nevertheless, tramways were only an expensive and temperamental tool, and their services were required only as long as the mines produced. Given the vagaries of underground mining, it should be no surprise that many tramways, so carefully constructed, were abandoned or dismantled within an incredibly short time. Others served a much longer life, although they too were subject to the ever-changing dynamics of western mining. In short, no matter how well built, aerial tramways could never be a permanent fixture. After 1920 numerous factors conspired to bring most of these spectacular structures to a standstill.

CHAPTER FIVE

DECLINE, OBSOLESCENCE, AND PRESERVATION

The 1920s witnessed the beginnings of a decline in the use of aerial tramways at western mine sites, a decline that has continued to a point where they have virtually disappeared from the contemporary landscape. Progress in mining technology as well as events beyond the industry's control have rendered, with few exceptions, aerial tramways an obsolete remnant of the past. The decline began as mining fell on hard times during the 1920s. This was compounded during the Depression when most of the surviving operations were forced to shut down or cut expenditures. World War II ended the mining of precious metals, cut out small operators, and saw scrap drives scoop up much of the surviving machinery. During the latter half of the twentieth century, mining became more mechanized with open pit mines, huge trucks and shovels, conveyor systems, and more efficient road-building equipment. A few operating aerial tramways still exist, but for the most part the historic survivors are now no more than rusting reminders of America's mining past.

Despite the fact that many mine operations lost money or went out of business during the 1920s, aerial tramways remained a popular tool. During the decade, Leschen & Sons, American Steel & Wire (Trenton), Broderick & Bascom, and Riblet remained active in the business. Each company offered a slightly different version of the double-rope tramway, provided free engineering designs, and helped with actual construction.[1] These companies were joined

Basic design of the Lawson tramway showing loaded and empty cars traveling on wire tracks. Courtesy Interstate Equipment Corporation.

just before the beginning of the decade by a new competitor with a radically different design, one that promoted automatic operation and the elimination of most labor expenses.

The new design was the brainchild of inventor William C. Lawson of Roanoke, Virginia. Lawson had founded the Interstate Equipment Corporation in 1903 and during the following decade he secured several patents for his "Loop-line Tramway." The system featured cars (much like mine cars) mounted with grooved wheels that rode on two parallel track cables, pulled by a traction cable attached directly to the bottom of the ore car. The car discharged its contents at the terminal and returned to the mine on a second line that ran either alongside or under the first one. Although the idea of a two-rail cable was not new (a tramway of this general design entered service in 1912 to link the Midget and Moon Anchor mines with the Wishbone Mill at Cripple Creek, Colorado), Lawson perfected a method of maintaining equal tension on both cables, thus preventing the shifting of loads and unequal wear.[2]

In the early 1920s Interstate acquired a small Virginia company called Continental Tramway. Under the direction of President Edward J. Grassman and design engineer Frank Lawrence, this small company had manufactured and installed tramways in the eastern United States at least since 1910. After Interstate acquired Continental, Lawson established an office at 25 Church Street, New York City, where the responsibility for design, engineering, erection, and sales was centered. Manufacturing remained at the Continental plant in Roanoke. As soon as these arrangements were completed, Interstate began to market the Lawson system in the western United States under the trade name "Automatic Aerial Tramway."[3]

While most of the Lawson tramways continued to be installed east of the Mississippi, particularly at Appalachian coal mines, a number of the devices went to western mining operations during the 1920s. One was erected for the Nephi Plaster and Manufacturing Company at Nephi, Utah. Designed to transport gypsum to a railroad siding some 9,500 feet from the mine "regardless of weather conditions, over hog backs, chasms and rough country," it could be operated by one man. The gypsum mine owned by the Blue Diamond Company some thirty miles southwest of Las Vegas, Nevada, which reportedly spent over $1 million to develop its property, also installed a Lawson tramway in 1925. A brochure declared it "one of the most daring undertakings in the history of plaster manufacturing." That same year, Interstate put in two separate tramways at a rock quarry operated by the Basalt Rock Company at Napa, California.[4]

Despite a marked reduction in new mining ventures during the 1920s, several conventional tramways also entered service. In January 1926, for example, the Trenton Company completed a tramway "as modern in every detail as money and engineering can make it." Operated by Park-Utah Consolidated Mines at Park City, Utah, the 4,700-foot unit offered a number of advanced features. All towers were fabricated from heavy steel set on concrete footings. The drive provided two speeds—250 and 500 feet per minute—and all ore handling was fully automatic. Incoming supplies such a lumber, coal, and miscellaneous freight also rode the tramway. Because the mine produced various grades of ore plus zinc and lead concentrates, a system of conveyor

Discharge terminal of a Lawson design tramway at an unidentified location in Salt Lake County, Utah, about 1920. Note use of tractor and ore wagons used to haul ore to processing. Courtesy Interstate Equipment Corporation.

belts and bins separated the diverse products so that each could reach its intended destination via the same tramway.[5]

A year later, the American Metal Company leased the Pecos zinc-lead mines near Glorieta, New Mexico. To recover mineral deposits known to exist since 1882, American Metal invested over $2 million to construct a 600-ton flotation mill and a 3,000-kilowatt power plant at Alamitos, near the community of Pecos. Because the mine lay some twelve miles distant, the company employed a standard Riblet tramway requiring one angle station. Utilizing a 100-horsepower electric motor, the line operated at a speed of 500 feet per minute. With four hundred buckets delivering ore to the mill, the tramway had a rated capacity of 60 tons per hour.[6]

Meanwhile, the Bunker Hill and Sullivan Mining Company of Idaho, which had grown during the early century into a "world-class" mining operation by acquiring mines both inside and outside the Coeur d'Alene district, decided to erect an aerial tramway at its Sidney operation. Rather than hire one of the major companies to construct the 15,650-foot line, Bunker Hill managers opted to contract with the Painter Tramway Company to provide the design and machinery. Track cables came from Roebling's and the traction rope from American Steel & Wire. A detailed cost estimate reveals that for the 210-day construction period, it would cost $49,453 for labor, $47,102 for materials, and $14,673 for transportation. After calculating for a loss caused by a small fire in April 1928, the final cost was expected to be $114,494. The finished structure was to feature twenty standard towers, one 104-foot tower, and two tension stations.[7]

Two of the more interesting tramways to enter service during the 1920s were operated by the Santa Catalina Island Company, owned by William Wrigley Jr. Located in the Pacific Ocean some twenty-six miles off the coast of southern California, Catalina Island possessed a number of natural resources, including three mines, two of which operated tramways. A producer of lead and zinc ore, the Renton Mine at the island's south end operated a 3,700-foot aerial tramway in order to transport the raw minerals to barges for transfer to the mainland. The much larger Black Jack Mine in the middle of the island produced high-grade argentiferous galena and zinc. To process this ore, the Catalina Island Company erected a 100-ton flotation mill at White's Landing. Because the mine was located on a hillside some 1,540 feet above the mill, the company built a 9,900-foot aerial tramway to link the two sites. The double-rope tramway cost $30,000 to install and utilized 800-pound-capacity buckets.[8]

By far the most impressive tramway of the decade (and one of the last built before the Depression) went up in 1929. Organized by Charles Chase in 1926, the Shenandoah-Dives Mining Company (captialized at $3.5 million) bought several old properties in Arastra Basin outside of Silverton, Colorado. To service these mines, Chase constructed the 300-ton Mayflower Mill along the Animas River near the old Silver Lake complex. Although several abandoned tramways remained in the area, Chase opted to construct a totally new system. Employing F. C. Carstarphen of Denver to design the 10,000-foot tramway, Chase wanted everything to be world-class. As a consequence, all the structures (two tension stations, one rail station, and eleven towers) were steel. The towers used a square frame design that increased "the number of points of support, thereby diminishing the magnitude of the deflection angles." The one-and-a-quarter-inch track cables stretched tight under 35,000 pounds of tension. Buckets consisting mostly of 1,450-pound-capacity units rode on four-wheel carriages that moved at a speed of 500 feet per minute. These features enabled the tramway to transport 55 tons of ore per hour, operated by five men per shift—"two at the loading terminal, two at the discharge terminal, and a foreman." Despite some construction delays, the Shenandoah-Dives tramway transported its first ore on February 10, 1930.[9]

Mayflower Mill (now a museum) and terminal of the 1929 Shenandoah-Dives tramway near Silverton, Colorado. Author's photo.

The Shenandoah-Dives also operated a "stub" tramway at the upper terminal. This short line carried supplies such as food and fuel to the bunkhouse as well as moving heavy construction materials up to the main mine entrance. The stub tram ran under the supervision of a colorful character named Bill Smith, who was locally famous for his temper. An old-timer recalls that "whenever the boys in the tram house would give Bill repeated bells for him to bring the bucket up and there wasn't any bucket, he'd get aggravated. If they kept it up, he'd get mad, and then he'd run down the hill with an axe in his hand. The boys would lock the doors, and it would be a while before anyone would signal the stub tram again." The Shenandoah-Dives tramway also became famous for a huge masonry "slide splitter" built in 1938 to deflect avalanches from a vulnerable tower. This stone and cement structure remains standing today as a monument to the quality workmanship of the time. Many of its component materials came up to the site on the tramway itself.[10]

One final construction effort took place before the full impact of the Depression could be felt. In early 1930 Newmont Mining Company, which

One of the modern steel towers on the 1929 Shenandoah-Dives tramway at Silverton, Colorado. This tramway remains largely intact today. Author's photo.

Shenandoah-Dives tramway high-capacity ore bucket at the unloading dock of the Mayflower Mill. Author's photo.

controlled the Empire and North Star properties in California's Grass Valley gold district, elected to build a 6,000-foot tramway connecting the North Star Mine with the eighty-stamp Empire Mill. This system entered service in early 1931, but it is unknown how long it operated.[11]

The stock market crash and subsequent depression took a heavy toll on western mining operations. Marginal companies closed down and many of the survivors operated only part-time. Most tramways fell silent. A few, however, managed to remain in business. Utah's Highland Boy tramway at Bingham Canyon carried on through the Depression by operating at 50 percent capacity three days per week. The Shenandoah-Dives did even better. Operating almost continuously, it moved approximately 1,800,000 tons of ore during the 1930s. Despite having to make financial cutbacks, General Manager Charles Chase managed to keep most of his miners employed and make improvements in his plant, helped in part by the government's raising the price of gold to thirty-five dollars an ounce in 1935. At several locations in the San Juans, defunct mines were leased during the 1930s in order to reclaim tailings and do placer work. Using the flotation process, such noted properties as the Camp Bird, Smuggler-Union, and Sunnyside experienced a brief revival. At the Smuggler-Union and perhaps others, the tramway was returned to service.[12]

Although the Depression forced many tramways out of business, mining engineers continued to improve the technology. American Steel & Wire pretty much remained the market leader and aggressively sought customers. Its 1935 catalog offered a variety of different aerial wire-rope tramway systems using both creosoted wood and steel structures. Because of its long history in the business, American Steel & Wire boasted that it possessed such a great store of valuable engineering and practical experience that its engineers offered the most advanced tramway structures available. In another direction, Fred C. Carstarphen, a consulting engineer from Denver who specialized in tramway design, began to suggest that the old idea of preferring a straight-line system was outdated. By utilizing a series of angle stations (which had now been perfected), the route of the tramway could be made more effective, enabling the line to skirt steep mountain walls, operate with milder gradients, and reduce construction and operating costs. In 1930 Carstarphen also published a comparison between truck and tramway costs. His figures showed that except for short distances and small loads trucks offered no competition. "This is because," he noted, "the tramway terminals are a fixed charge irrespective of the length of the line and greatly exceed the cost of a light-weight truck." Meanwhile, the Riblet Company continued to improve long-distance tramway technology. Working largely in South America, the Spokane-based company constructed tramways over distances as great as thirty miles. One system, built for the Northern Peru Mining and Smelting Company, included four terminals, four control stations, two angle stations, forty tension stations, and 308 towers over its 30.79-mile length.[13] Should the need arise Riblet stood ready to construct long-distance tramways in the United States, although orders were few.

Another innovation that grew out of materials technology was the recreational or tourist tramway. Although unrelated directly to mining, the increasing popularity of passenger lifts helped keep the tramway business alive during the Depression. Tramways capable of carrying passengers, of course, had been around for years, but by the 1920s European tramway builders, such as Adolf Bleichert & Company, were constructing double-rope gondola systems specifically to reach scenic points difficult to access "by the great majority of travelers and folk bent on pleasure." During the 1930s the chair lift emerged specifically for the purpose of transporting skiers to the top of favorite slopes "that would otherwise be closed to the great majority of sports lovers." One of the first such operations, at Cement Creek in Colorado, simply converted an existing mine tram to recreational use by replacing the buckets with chairs. Despite major safety issues, this line remained in use for a number of years. In 1937 the Union Pacific Railroad built a chair lift specifically for skiing at Sun Valley, Idaho. So popular was this installation that six more lifts entered service at Sun Valley within the next few years. All these devices used the same basic cable technology (either single- or double-rope) that had proved so successful in materials handling. After World War II these recreational lifts became so common that they assumed a leadership position in the industry through their mechanical improvements.[14] Ironically, aerial tramways helped rejuvenate many towns, especially in Colorado and Utah, that were once bustling mining centers. At the end of the twentieth century, places such as Aspen, Telluride, Park City, and Alta are again alive with activity because of the skiing industry, which depends heavily on lifts that reach into the same mountains where mine tramways once ran.

World War II changed mining even more. Mines unable to produce war-related materials were shut down, while those producing base metals such as copper, lead, and zinc did well. Such companies as the Shenandoah-Dives were able to turn out large quantities of strategic minerals. After a short post-war boom, however, mining fortunes turned downward again. Mines closed by the war found it too expensive to reopen and even those that stayed open fell on hard times. By the 1950s many of the prewar mines were out of business. In 1953, for example, the Shenandoah-Dives closed down, permanently idling its tramway (except for a brief revival during the 1957 filming of *Night Passage*, a western adventure film starring Jimmy Stewart that used the tramway for a daring shoot-out).[15]

Those mines that survived increasingly converted to newer technology. As high-tonnage, low-grade workings became the norm, tramway systems could not financially compete with other forms of transportation. The development of open pits, haulage tunnels, long-distance conveyors, and high-capacity trucks made one-ton aerial tramway buckets obsolete. At underground workings, it sometimes proved cheaper in the long run to dig a tunnel in at the mill level and raise up under the existing workings. This happened at the Pine Creek tungsten operation near Bishop, California. To tap the mine's rich high-altitude deposits, a two-mile tramway had originally been constructed in 1937. After World War II, as ore grades diminished, owner Union Carbide Corporation decided to tunnel into the mountain directly from the mill. Completed in 1970, the tunnel "was given the name Easy Go because it would eliminate the tramway and the daily commute by miners." Trains were used to bring the ore directly out of the tunnel.[16] As a consequence of such improvements and large-scale mining techniques, tramways (which were always the most economical when transporting high-grade ores) became less useful and more expensive.

Although western mine tramways were largely surpassed by other technology after the war, a few notable exceptions could be found. In Arizona, three new aerial systems were constructed after 1950, two of which were located in the Grand Canyon. The Orphan Mine, a leading high-grade uranium mine on the South Rim of the canyon, proved to be one of Arizona's most controversial ventures. Claims to the Orphan Lode, located in Grand Canyon National Park just west of Grand Canyon Village, date back to the 1890s when Daniel L. "Pops" Hogan discovered an outcropping of unusual minerals about 1,100 feet below the rim. In 1906, just before the federal government began protecting the canyon, Hogan secured a patent for his claim, which he took to be rich in copper because of its green-colored ores. Despite Hogan's failure to make anything of his mine, he held onto the property until 1947. By this time the Orphan Mine represented the only privately held land within park boundaries and was coveted by the Park Service, which viewed it as an ugly intrusion on the park. Nevertheless, Hogan sold the claim to a woman from Prescott, Arizona, who used the land on the rim to construct a tourist lodge. Meanwhile, assays showed that the mine's bright green ore actually contained high-grade uranium.[17]

In light of this development, the Orphan was sold to the Golden Crown Mining Company in 1953. Hoping to cash in on the booming demand for uranium, Golden Crown (a subsidiary of Western Gold & Uranium) sent out geologists to determine the most practical way to bring ore 1,100 feet to the canyon's rim. In 1955 they decided to build an aerial tramway rather than service the mine by Hogan's steep trail and series of ladders. After an early tramway proved unworkable, materials for a second line were acquired from the Riblet Company and installed by Western Gold & Uranium crews at a total cost of $61,800. Actual construction presented some unique challenges. Because of the steep canyon walls, the towers were set in place by workers strapped to bosun's chairs suspended by ropes. When completed, the tramway stretched a total distance of 1,800 feet. Supported by several intermediate towers and two terminal structures, the line descended at a thirty-seven-degree angle for the first 1,000 feet, then at fifty-seven degrees for the final 800 feet. Ore deposited atop the rim was trucked to a mill at Tuba City, Arizona, for processing. This tramway entered service in April 1956.[18]

Operating opposite of most mine tramways, the Orphan system carried ore uphill and supplies downward. Because of the rugged terrain, miners also had to ride the buckets to work. Although the mine was one of the richest in the United States, the Park Service wanted what it saw as an eyesore removed from the middle of a pristine area visited by millions of tourists each year. Local politicians, however, viewed the mine as a key factor in the survival of northern Arizona's uranium industry and sought to secure access to adjacent ore deposits under park land. Meanwhile, the tramway proved inefficient and expensive to operate. The two-bucket jig-back system limited the amount of ore that could be carried to the rim, pushing costs to about twenty-five dollars per ton. To counter this problem, the owners sunk a vertical shaft from the rim directly to the ore body in 1959. With this, the tramway became obsolete for hauling ore, although it continued to be used for several more years to transport men and supplies. In 1969 the mine ceased operations, largely because of Park Service opposition, a decline in uranium prices, and the fact that ore now had to be carried to Colorado for processing. The Orphan claim was finally turned over to the Park Service in 1987 and portions can still be observed near the rim.[19]

The other Grand Canyon tramway, also built in the late 1950s, became famous as a major boondoggle. This system was created to exploit the Bat Cave, a natural cavern located in the western part of the canyon, about 800 feet above the Colorado River. The cave derived its name from deposits of bat guano, which in the 1950s was being touted as "nature's most perfect plant food." In 1958 the U.S. Guano Company leased the remote cave in order to extract the estimated 200,000 tons of guano deposited over the centuries by millions of bats. After unsuccessfully experimenting with the use of boats and aircraft to remove the guano, the company contracted with U.S. Steel's Western Steel Division to construct a 10,000-foot modern double jig-back system using two large ore carriers. Again, construction presented challenges, including how to get the initial cable across the canyon, a feat finally accomplished by helicopter. When completed, one span—from the south rim to a tower near the river—covered a mile and a half, giving the tramway the longest span across a canyon in the United States. Unfortunately, the "Bat Cable" experienced trouble from the start. During the first year, cables snapped twice, crashing to the canyon floor. Once operations began, it was discovered that the cave contained only 1,000 tons of guano, and within a year the mine played out. As one critic noted: "the Company has spent $3.5 million to salvage 1,000 tons of guano which sold for 69 cents a pound." Shortly after the mine closed a military jet clipped one of the lines, which dropped into the canyon. The remaining cable was used only one final time to make the 1959 film *Edge of Eternity*, starring Cornell Wilde, which utilized the tramway in some hair-raising stunts. Although a costly failure, U.S. Steel considered the project worthwhile because "we got a million dollars worth of free advertising doing the job."[20]

One of the most recent tramways went up at Morenci, Arizona, in 1970. Built to connect a limestone quarry east of the San Francisco River with the

U.S. Guano Company jig-back tramway at Grand Canyon, Arizona, in 1959. Note that the bucket is fitted with a passenger section. Miners could only reach the mine by tramway. Courtesy Mohave County Historical Society, negative 5486.

Phelps Dodge smelter at Morenci, the four-mile tramway was built by the Interstate Equipment Corporation using a modernized version of the Lawson system it first introduced in the 1910s. Employing several high towers, cars

U.S. Guano Company tramway looking at the South Rim discharge terminal for the "Bat Cable," Grand Canyon, Arizona, in 1959. Courtesy Mohave County Historical Society, negative 9261.

holding 2,000 pounds of limestone ran on two parallel cables during the slow trip from one end to the other. At the unloading terminal, "each car was flipped over on its back, the rock would fall out, and the car would ride back to the quarry upside down for another load." This tramway operated for less than a decade before a better source of limestone was located. It then remained idle until February 16, 1997, when Phelps Dodge blasted it down with dynamite.[21]

Although tramways in western mining districts are now rare, they have not entirely disappeared. During the last two decades of the twentieth century, industry journals have stressed a number of technical improvements that may make tramways practical in some instances. As one article concluded, "The American mining industry is rediscovering aerial cableway technology." One factor in the renewed interest has been improved reliability due to advances derived from ski-lift technology, which has continued to advance. Nonetheless, mine tramways are advised only under certain circumstances: (1) when yearly capacity will exceed 1 million tons, (2) where the line will be used for a minimum of three years, (3) where grades are greater than 12 percent, and (4) where the length is over a mile. While most current installations are overseas, both the Interstate Equipment Corporation and Riblet continue to build or rebuild tramways in the United States and Canada. Indeed, during the 1990s Interstate completed two projects for the Golden Bear (gold mining) Corporation at Dease Lake, British Columbia, near the site of a 15,370-foot tramway the company installed in 1974 for the Cassair Asbestos Corporation. Aerial tramways will never be as popular as they once where, but as one writer concluded in 1985, "The aerial cableway is the solution when in frustration you say 'If I could only fly this material to the tipple.' The cableway is more like flying than any other system."[22] This same sentiment rang true in the 1890s.

Given the historical significance of aerial tramways to the development of western mining, it is discouraging that so little has been accomplished in the realm of historic preservation. Most of the tramways were abandoned or stripped of their machinery well before their historical significance could be appreciated. As a consequence, much of their physical presence is gone. Nevertheless, at many locations a visitor can still find the remains of towers, terminals, and cable. Retired but basically intact tramways are rare. Unfortunately, little is being done to protect remaining lines so that future generations may appreciate the importance of these devices during the heyday of western mining. Like so much of mining history, old tramways are dangerous and would be terribly expensive to maintain, making it impractical to do much for the survivors. Luckily, a few specimens are available to visitors interested in these fascinating pieces of machinery.

Perhaps the only remaining intact Hallidie single-rope tramway can be found at the Le Roi Mine and Museum at Rossland, British Columbia. Although short and somewhat overgrown with trees, one can get a clear view of the towers, cables, and permanently attached buckets. Certainly one of the most impressive survivors is the 1929 Shendadoah-Dives line near Silverton,

One of the surviving tramway towers for the Mountain Hero Mine, Yukon Territory, Canada. Many such structures remain across the West as reminders of an earlier era. Courtesy Murray Lundberg Collection, Yukon Archives, 95/105 #231.

Colorado, which ceased operations in 1953. In 1996 this modern tramway was acquired by the San Juan County Historical Society along with the Mayflower Mill. The mill has since been opened for tours, which include an inspection of the tramway terminal. Although this system is no longer capable

Ruins of the Little Nation Mill and tramway terminal at Howardsville, Colorado. Note that a new roof has been placed on the structure to help preserve what is left of the building. Although the cables remain in place, there is some discussion of taking them down for safety purposes. Author's photo.

of operation and the cables are secured in concrete, one can inspect almost all aspects of the operation. From the mill the cables and buckets can be seen stretching some two miles up Arastra Basin. Another largely intact tramway is located at Pioche, Nevada. Built for the Pioche Mining Company during the 1920s, this double-rope system with steel towers and automatic dumping buckets once delivered ore from the mines of Treasure Hill to Godby's Mill. While the wooden loading terminal has deteriorated significantly, the mechanical works and bull wheel are clearly visible.

Several tramways also fall under the jurisdiction of the National Park Service, whose policy might best be described as ranging from benign neglect to arrested decay. The most complete system is the series of tramways built at the Kennecott Mines, a National Historical Landmark located in Wrangell–St. Elias National Park, Alaska. This remote copper-mining and -milling complex, built between 1905 and 1911, operated several lengthy tramways. The mines closed in 1938, but most of the structures, including the tramways,

Remains of the Gold Prince tramway as they stand in 2001 in Mastodon Gulch near Animas Forks, Colorado. Author's photo.

were left intact. Now under the care of the Park Service, little has been done in the way of preservation. Presently, the wooden towers are decaying and are barely able to support the cables. Fearing the possibility of a sudden collapse, there has been talk of lessening cable tension in an effort to stabilize the tramways, but as of this writing nothing has been done. The Keane Wonder tramway in Death Valley is also available to be seen. In dilapidated condition, with the cables down in many places, most of the tramway structures nevertheless remain and can be inspected by a relatively short hike. As part of Death Valley National Park, the site is somewhat protected from vandalism, but beyond putting some money into monitoring and stabilizing the remnants little is being done to prevent further deterioration.[23] As with all historic mining structures, I urge the reader to see what is left of the tramways while you can. Every year the weather and vandals reduce more of these once proud machines to unrecognizable rubble. Like the men who worked on them, they will all be gone sooner than we would like.

NOTES

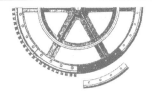

INTRODUCTION

1. *Times-Gazette* (Redwood City, California), March 5, 1898.
2. T. A. Rickard, "Across the San Juan Mountains," *The Engineering and Mining Journal* (hereafter *E&MJ*) (August 22, 1903): 26 [This article was one of a series printed in the *E&MJ*. All of them were subsequently published in Rickard's *Across the San Juan Mountains* (New York: The Engineering and Mining Journal, 1903)]; "Aerial Tramways and Mining," *The Mining and Metallurgical Journal* (November 5, 1908): 1.
3. Zbignieu Schneigert, *Aerial Tramways and Funicular Railways* (Warsaw, Poland: Pergamon Press, 1966), 1–4; Alexander J. Wallace-Taylor, *Aerial or Wire Rope-Ways: Their Construction and Management* (London: Crosby Lockwood and Sons, 1911), 1–2; "An Anticipated Invention," *Mining and Scientific Press* (hereafter *M&SP*) (January 31, 1874): 65.
4. Schneigert, *Aerial Tramways*, 3–4; Lynn R. Bailey, *Supplying the Mining World: The Mining Equipment Manufacturers of San Francisco, 1850–1900* (Tucson, AZ: Westernlore Press, 1996), 115; "Origins of Wire Ropes," *M&SP* (August 2, 1873): 67; "Facts about Wire Rope," *M&SP* (August 16, 1884): 102. A description of early wire rope making can be found in *Wire Rope and Its Applications* (St. Louis: Broderick & Bascom Rope Company, 1939), 3–4.
5. Bailey, *Supplying the Mining World*, 116; "Origin of Wire Ropes," *M&SP* (August 2, 1873): 67; "Facts about Wire Rope," *M&SP* (August 16, 1884): 102.

6. D. B. Steinman, *The Builders of the Bridge: The Story of John Roebling and His Son* (New York: Harcourt, Brace and Company, 1945), 61–64; *Roebling Wire Rope and Wire, Pacific Coast Edition* (Trenton, NJ: John A Roebling's Sons Company, 1931), 1–2.

7. Steinman, *The Builders of the Bridge*, 65–66.

8. Ibid., 135–136; Edward C. Mack, *Peter Cooper, Citizen of New York* (New York: Duell, Solan and Pearce, 1949), 206; Rossiter W. Raymond, *Peter Cooper* (reprint of 1901 ed., Freeport, NY: Books for Libraries Press, 1972), vol. II, 532–534; *Acts of the 71st Legislature of the State of New Jersey* (Trenton: Phillips and Boswell, 1847), 61–64.

9. "Endless Wire Rope Tramways," *E&MJ* (July 9, 1872): 26; Raymond, *Peter Cooper*, 32.

10. "Wire Tramways," *E&MJ* (June 20, 1871): 385; "Endless Wire Rope Tramways," *E&MJ* (July 9, 1872): 26; Mack, *Peter Cooper*, 210.

11. "Wire Tramways," *E&MJ* (August 13, 1872): 106; Schneigert, *Aerial Tramways*, 4; various notes, Northern Mine Research Society, Keighley, Yorkshire, England.

12. Wallis-Taylor, *Aerial or Wire Rope-Ways*, 1–2.

CHAPTER 1

1. Rodman W. Paul, *Mining Frontiers of the Far West, 1848–1880* (New York: Holt, Rinehart and Winston, 1963), 12–36; Otis E. Young Jr., *Western Mining* (Norman: University of Oklahoma Press, 1970), 102–124.

2. Paul, *Mining Frontiers*, 68.

3. David F. Myrick, *Railroads of Nevada and Eastern California, vol. I, The Northern Roads* (Berkeley, CA: Howell-North Books, 1962), 136–138.

4. Paul, *Mining Frontiers*, 102–105.

5. Bailey, *Supplying the Mining World*, vii.

6. Edgar M. Kahn, *Andrew Smith Hallidie: A Tribute to a Pioneer California Industrialist* (San Francisco: privately published, 1953), 1–13; Edgar M. Kahn, "Andrew Smith Hallidie," *California Historical Society Quarterly* 19 (June 1940): 144–146; "Facts About Wire Rope," *M&SP* (August 16, 1884): 102. For a brief overview of tramway history, see Robert A. Trennert, "From Gold Ore to Bat Guano: Aerial Mine Tramways in the West," *The Mining History Journal* 4 (1997): 3–113.

7. "Hallidie's Improved Suspension Bridge," *M&SP* (October 23, 1869): 265; Bailey, *Supplying the Mining World*, 122; Kahn, "Andrew Smith Hallidie," 152.

8. "The Wire Tramway," *M&SP* (August 7, 1869): 82.

9. [Andrew S. Hallidie], *Transportation of Ore and Other Material by Means of Endless Traveling Wire Ropes* (San Francisco: C. A. Murdock & Co., 1878), 1–25. See also "Hallidie's Endless Wire Rope Tramway," *E&MJ* (July 2, 1872): 2; Edward B. Durham, "Aerial Tramways and Cableways," in Robert Peele, ed., *Mining Engineer's Handbook* (New York: John Wiley & Sons, 1927), vol. II, 1787–1788; and O. H. Metzger, "Aerial Tramways in the Metal-Mining Industry, Part I," U.S. Bureau of Mines, Information Circular #6948 (September 1937), 3–4.

10. [Hallidie], *Transportation of Ore*, 4; "Hallidie's Improved Suspension Bridge," *M&SP* (August 16, 1869): 265; United States Patents #100,140 (February 18, 1871) and #110,971 (January 17, 1871).

11. "Hallidie's Endless Wire Rope-Way," *M&SP* (February 18, 1871): 104–105.

12. Advertisement, *M&SP* (March 11, 1871): 160.

13. Schneigert, *Aerial Tramways*, 4; English Patent #2281 (July 20, 1868) and miscellaneous information from the records of the Northern Mine Research Society.

14. For articles describing Hodgson's tramway, see "Wire Tramway at White Pine," *M&SP* (August 20, 1870): 128; and "Wire Tramway," *M&SP* (September 10, 1870): 185.

15. Paul, *Mining Frontiers*, 106–107; W. Turrentine Jackson, *Treasure Hill: Portrait of a Mining Camp* (Tucson: The University of Arizona Press, 1963), 5–125.

16. Melville Atwood to Thomas Phillpotts, July 28, 1870, and Frederick F. Haggard to Phillpotts, August 11, 1870, Thomas Phillpotts Correspondence, 1870–1873, William Miles Read Papers, the Bancroft Library, Berkeley, California; Jackson, *Treasure Hill*, 122, 176–177; *M&SP* (March 21, 1870): 197 and (April 30, 1870): 285.

17. Jackson, *Treasure Hill*, 177; *M&SP* (December 3, 1870): 380, (December 17, 1870): 413, and (December 24, 1870): 438.

18. "A Visit to Treasure Hill," *M&SP* (October 7, 1871): 211; *M&SP* (January 28, 1871): 53, (March 18, 1871): 165, (April 15, 1871): 229, (May 13, 1871): 293, and (May 27, 1871): 325.

19. C. J. Bulkley to Phillpotts, January 6, 1872, Phillpotts Correspondence; "A Visit to Treasure Hill," *M&SP* (October 7, 1871): 211; *M&SP* (December 23, 1871): 385.

20. Bulkley to Phillpotts, April 15, [1873], Phillpotts Correspondence; Jackson, *Treasure Hill*, 178; *M&SP* (December 23, 1871): 385, (January 13, 1872): 20. Hodgson's company went out of business in 1873 and was declared bankrupt in January 1874, although others continued to build tramways of his design, mostly in Europe.

21. United States Patents #115,309 (May 30, 1871), #115,310 (May 30, 1871), and #121,776 (December 12, 1871).

22. *Territorial Enterprise* (Virginia City, Nevada), January 23, 1870; *Report of the Mineralogist of the State of Nevada for the Years 1869 and 1870* (Carson City: Charles L. Perkins, 1871), 100–101; Myron Angel, ed., *History of Nevada* (Oakland: Thompson and West, 1881), 485.

23. *Territorial Enterprise*, October 26, 1871; *M&SP* (February 3, 1872): 69, (July 20, 1872): 44; [Hallidie], *Transportation of Ore*, 6.

24. *M&SP* (July 20, 1872): 44; "Air Transportation," *E&MJ* (August 20, 1872): 122.

25. Duane A. Smith, "'Where a Bird Could Hardly Obtain a Footing': George Armstrong Custer and the Stevens Mine," *Colorado Heritage* 17 (Spring 1997): 25–35; "Air Transportation," *E&MJ* (August 20, 1872): 122.

26. "An Endless Wire Rope-Way for Little Cottonwood Canyon," *M&SP* (May 31, 1873): 344; "Alta City, Little Cottonwood, Utah," *M&SP* (November 22, 1873): 329.

27. [Hallidie], *Transportation of Ore*, 6–7.

28. "An Endless Wire Rope-Way for Little Cottonwood Canyon," *M&SP* (May 31, 1873): 344.

29. "History of the Emma Mine," *M&SP* (January 10, 1874): 18; "Cottonwood Canon Mines," *M&SP* (January 31, 1874): 66. For analysis of the Emma Mine swindle, see W. Turrnetine Jackson, "The Infamous Emma Mine: A British Interest in the Little Cottonwood District, Utah Territory," *Utah Historical Quarterly* 23 (October 1955): 339–362; and Clark C. Spence, *British Investments and the American Mining Frontier, 1860–1901* (Ithaca, NY: Cornell University Press, 1958), 139–182.

30. [Hallidie], *Transportation of Ore*, 7–8; "Manufacture of Wire Rope in San Francisco," *E&MJ* (April 10, 1875): 243; Articles of Incorporation, California Wire Works, April 20, 1882, California State Archives, Sacramento; *M&SP* (June 10, 1882): 383; "The California Wire Works," *M&SP* (August 2, 1884): 69; Bailey, *Supplying the Mining World*, 122–123.

31. "Endless Wire Rope-Way," *M&SP* (September 5, 1874): 145; [Hallidie], *Transportation of Ore*, 3, 6–8.

32. Various advertisements, *M&SP* (1884); "The Denver Exposition," *M&SP* (September 15, 1883): 162; Allen Nossaman, *Many More Mountains, vol. III, Rails into Silverton* (Denver: Sundance Books, 1998), 310–313.

33. "The Denver Exposition," *M&SP* (September 15, 1883): 162.

34. "Huson's Wire-Rope Tramway for Mines," *M&SP* (January 1, 1887): 1. See also United States Patent #312,274 (February 17, 1885) for a description on Huson's "Device for Loading and Unloading Tramways."

35. Bailey, *Supplying the Mining World*, 88; *Huson's Patent Automatic Wire Rope Tramway* (Denver: C. W. Badgely & Co. [1893]), 1–5, Olcott Papers, New York Historical Society.

36. Badgely catalog, 18; *Silverton Weekly Miner*, January 8, 1897.

37. Badgely catalog, 19; Mallory Hope Ferrell, *Silver San Juan: The Rio Grande Southern* (Boulder, CO: Pruett Publishing Company, 1973), 86; Russ Coleman, Dell A. McCoy, and William A. Graves, *The RGS Story* (Denver: Sundance Books, 1993, 1994), vol. III, 434, vol. IV, 141.

38. "Wire Rope Tramway at the San Juan Mines," *Engineering News* (February 16, 1893): 146; "New Clips of Wire Ropeways, " *E&MJ* (February 17, 1900): 202.

39. "Hallidie Ropeway at Halls Mines, Nelson, B.C.," *M&SP* (February 1, 1896): 84–85; Garnet Basque, *West Kootenay: The Pioneer Years* (Langley, B.C.: Sunfire Publications, 1990), 45–47.

40. "The Tramway at El Dorado Mine, Utah," *E&MJ* (April 13, 1901): 461; "Andrew S. Hallidie," *M&SP* (April 20, 1900): 462.

CHAPTER 2

1. Stephen de Zomdoria, "Aerial Tramways," *The Iron Age* (September 24, 1903): 30–31.

2. Ibid.; *Wire Rope Tramways for Economical Transportation: Bleichert Patent System* (New York: Cooper, Hewitt & Company, 1888), 3–6; Edward B. Durham, "Aerial Tramways and Cableways," in Peele, *Mining Engineer's Handbook*, vol. II, 1787.

3. Schneigert, *Aerial Tramways*, 5–6; J. Pohlig, "Aerial Wire Ropeways," *Transactions of the American Institute of Mining Engineers* 19 (1891): 761; Durham, "Aerial Tramways and Cableways," vol. II, 1753; Wallis-Taylor, *Aerial or Wire Rope-*

Ways, 9–10; "Relative Economy of Different Systems of Wire-Rope Convey-ance," *E&MJ* (February 14, 1880): 115–116.

4. *Wire Rope Tramways for Economical Transportation*, 8–9; "Relative Economy of Different Systems of Wire-Rope Conveyance," *E&MJ* (February 14, 1880): 115–116; Pohlig, "Aerial Wire Ropeways," 760–761.

5. U.S. Patents #380,982 and #380,983 (April 10, 1888); "The Bleichert System of Wire Rope Tramways," *E&MJ* (May 19, 1888): 361; *Wire Rope Tramways for Economical Transportation*, 5–8.

6. *Wire Rope Tramways for Economical Transportation*, 12–16.

7. "Contracting Notes," *E&MJ* (June 30, 1888): 478; *E&MJ* advertising supple-ment (December 29, 1888): xx; "The Bleichert Cable Transportation System," *Engineering News* (September 7, 1889): 225; *Wire Rope Transportation in All Its Branches* (Trenton, NJ: Trenton Iron Company, 1896), 30–31.

8. Robert Wayne Smith, *The Coeur d'Alene Mining War of 1892: A Case Study of an Industrial Dispute* (Corvales: Oregon State University Press, 1961), 3, 8, 24; *Wire Rope Transportation*, 9, 15.

9. *Spokane Spokesman*, December 24, 1891; *Spokane Review*, September 24, 26, and November 3, 1891.

10. V. M. Clement to John H. Hammond, January 15, 1892, Manager's Records, 1892–1901, Bunker Hill Company Records (MG 367), University of Idaho Library.

11. *Wire Rope Transportation*, 11–13, 15–17; Manager's Report, Bunker Hill and Sullivan Mining and Concentrating Co., May 31, 1892, Bunker Hill Company Records.

12. Clement to B. L. Eddie (telegrams), July 19, 20, 1892, F. W. Bradley to Clement (telegram), July 28, 1892, Clement to Hammond, September 1, 7, 11 (telegram), 12, 1892, Bradley to Hammond, December 17, 1892, Bunker Hill Company Records.

13. *Arizona Silver Belt* (Globe, Arizona), July 18 and August 8, 1891, and January 30, 1892.

14. "The Trenton Iron Company's Exhibit at Chicago," *E&MJ* (October 14, 1893): 394; *Wire Rope Transportation*, 30–31.

15. "A Wire Rope Tramway Operated at the Fair," *Scientific American* (October 7, 1893): 232–233; "The Trenton Iron Company's Tramway Exhibit at Chicago," *E&MJ* (October 14, 1893): 394; "Wire Ropeways at the Columbian Exposi-tion," *Engineering News* (November 9, 1893): 368–369.

16. "Wire Ropeways at the Columbian Exposition," *Engineering News* (November 9, 1893): 368–369.

17. *Wire Rope Transportation*, 39–44; "The 'Vulcan' Wire Ropeway," *M&SP* (March 21, 1891): 177, 185; "The Montgomery Wire-Rope Tramway," *E&MJ* (May 12, 1900): 563.

18. "A Long Span Wire Rope Tramway," *Engineering News* (April 2, 1892): 330; *Otto Aerial Tramways* (Chicago: Fraser & Chalmers, 1898), Arizona Collection, Arizona State University Library; "The Pohlig Universal Friction Grip," *E&MJ* (February 24, 1900): 232–233.

19. See various advertisements for Vulcan, Leschen, and Broderick & Bascom prod-ucts in the professional directory supplements of the *E&MJ* (December 30, 1899, June 29, 1901, June 28, 1902, December 31, 1903, December 29, 1904,

and December 26, 1904). For corporate information on A. Leschen & Sons, see Articles of Association, March 26, 1886, and May 30, 1903, Secretary of State Records, State of Missouri, Jefferson City.

20. *Spokesman Review* (Spokane, Washington), July 6, 1952; unidentified newspaper article, November 13, 1927, Cheney Cowles Museum, Spokane, Washington; John Fahey, "The Brothers Riblet," *Spokane Magazine* 4 (November 1980): 14–17.

21. Veronika Pellowski, *Silver, Lead and Hell: The Story of Sandon* (Sandon, B.C.: Prospector's Pick Publishing, 1992), 103; Fahey, "The Brothers Riblet," 14; William A. Barr, "Man against the Corporations," *Pacific Northwesterner* 31 (1987): 58–59. See *Official Gazette of the United States Patent Office*, November 17, 1903 (Washington, DC: Government Printing Office, 1904): 582–583, for a description of Riblet's grip patent.

22. For information on the Klondike gold rush and Chilkoot Pass, see Robert L. Spude, ed., *Chilkoot Trail: Historical Data* (Fairbanks: University of Alaska Press, 1980).

23. Frank Norris, "The Tramway Story," Karl Gurcke, ed., unpublished manuscript, Klondike Gold Rush National Historical Park, Skagway, Alaska; "Chilkoot Pass Route Tramways," Historic American Engineering Records AK-13 (blueprints), National Park Service, 1986–1987.

24. Ibid.

25. Ibid.; "A Cableway over Chilkoot Pass," *The Railroad Gazette* (December 24, 1897): 904–905; William Hewitt, "Across the Chilkoot Pass by Wire Cable," *Cassier's Magazine* (1898): 529–531.

26. Norris, "The Tramway Story"; Hewitt, "Across the Chilkoot Pass by Wire Cable," *Cassier's Magazine* (1898): 530, 538; "A Cableway over Chilkoot Pass," *The Railroad Gazette* (December 24, 1897): 905; *Times-Gazette*, March 5, 1898.

27. Norris, "The Tramway Story"; "Chilkoot Pass Route Tramways," Historic American Engineering Record, AK-13.

28. Rickard, *Across the San Juan Mountains*, 63–64.

29. "Wire-Rope Tramways," *E&MJ* (September 12, 1903): 398; "Wire-Rope Tramway," *E&MJ* (October 3, 1903): 514–515; Stephen de Zomdoria, "Rope Tramways," *E&MJ* (October 3, 1903): 513; Zomdoria, "Aerial Tramways," *The Iron Age* (September 24, 1903): 30–31.

30. California Wire Works, Articles of Incorporation, April 20, 1882, stamped "Charter forfeited Dec. 13, 1905, for failure to pay license tax of the year ending June 30, 1906," California State Archives.

31. *The Mining Catalog, Metal and Quarry Edition, 1921* (Pittsburgh: Keystone Consolidated Publishing Co., 1921), 244–245, 257–261; William Kent, *Kent's Mechanical Engineers Handbook* (New York: John Wiley & Sons, 1923), 1452; *A History of Trenton*, vol. II, 954. When the American Steel & Wire Company began using its own name is uncertain. However, a 1910 catalog, *The Bleichert System of Aerial Tramways*, prepared by William Hewitt (Trenton, NJ: The Trenton Iron Company), was overprinted with the American Steel & Wire name, suggesting that the change occurred at this time.

32. Quoted in Zomdoria, "Aerial Tramways," 31.

CHAPTER 3

1. Durham, "Aerial Tramways and Cableways," 1753; Rickard, *Across the San Juan Mountains*, 35; Edward G. Stoiber, "Excerpts: General Report upon the Silver Lake Mines, 1899," Duane Smith Collection. See also T. A. Rickard, "Mining at High Altitudes," *Cassier's Magazine* (October 1902): 695.

2. O. H. Metzger, "Aerial Tramways in the Metal-Mining Industry, Part 2. Construction and Operating Costs," U.S. Bureau of Mines, Information Circular #7095, February 1940, 3; Clement to Hammond, January 15, 1891, Bunker Hill Company Records.

3. "Proposal of A. Leshen & Sons Rope Company on an Aerial Wire Rope Tramway," St. Louis, Missouri, May 12, 1906, Ed Hunter Collection.

4. Durham, "Aerial Tramways and Cableways," 1754; Metzger, "Aerial Tramways, Part 2," 3–4.

5. Bunker Hill and Sullivan Mine, Minutes of the Board of Directors, October 20, 1892, Bunker Hill Company Records; Curtis H. Lindley, *A Treatise on the Law Relating to Mines and Mineral Lands*, 3d ed. (San Francisco: Bancroft-Whitney Co., 1914), vol. I, 570–571, 583–584; *People ex. rel. Aspen M. & S. Co. v. District Court, Colorado Reports* 11 (1887), 147–156.

6. M. A. Folsom to Harry L. Day, April 11, 1904, Hercules Mining Co. (RG 236), Miscellaneous Records, Tramway 1903–1904, University of Idaho Library.

7. Metzger, "Aerial Tramways, Part 2," 4.

8. Ibid., 5–6; *Riblet Aerial Tramways* (Spokane, WA: Shaw & Borden, 1930), 9.

9. Durham, "Aerial Tramways and Cableways," 1766–1768; "Long Cableways," *Engineering News* (April 16, 1892): 380; Hewitt, *Wire Rope Tramways*, 15.

10. C. M. Ballard, *Cutting Material Handling Costs* (Williamsport, PA: Williamsport Wire Rope Company, 1929), 90–91; Murray Lundberg, *Fractured Veins and Broken Dreams: Montana Mountain and the Windy Arm Stampede* (Whitehorse, Yukon Territory: Pathfinder Publications, 1996): 28, 84; Lundberg to author, June 2, 1998; W. Burns, "Report of the Preliminary Examination of the Gila Copper Sulphide Company, Christmas, Arizona, March 7, 1919," Christmas Mine File, Arizona Department of Mines and Mineral Resources; Stoiber, "Excerpts: General Report upon the Silver Lakes Mines, 1899."

11. "Track Rope for Tramways," *E&MJ* (September 25, 1915): 518; Durham, "Aerial Tramways and Cableways," 1763–1764; *Riblet Aerial Tramways*, 13.

12. *The Spokesman Review*, November 3, 1891.

13. "Novel Method of Cable Carrying," *M&SP* (August 7, 1897): 7. Clearly, this was not the first use of a mule train to transport tramway cable. In 1893 Hallidie had used the same basic technique to transport cable for a tramway in Baja California. See "Wire Rope Tramway at the San Juan Mines," *Engineering News* (February 16, 1893): 146.

14. "Unusual Feat of Transportation," *E&MJ* (June 15, 1907): 1159.

15. Schniegert, *Aerial Tramways*, 525–530.

16. Ballard, *Cutting Material Handling Costs*, 91.

17. "Turning Device for Tramway Track Cables," *E&MJ* (October 22, 1910): 801; "Track Ropes for Tramways," *E&MJ* (September 25, 1915): 518; Douglas Lay, "Aerial Tramways," *E&MJ* (February 28, 1920): 559, 561–562.

18. David F. Myrick, *Railroads of Arizona*, vol. III, *Clifton, Morenci and Metcalf, Rails and Copper Mines* (Glendale, CA: Howell-North Books, 1980), 724.

19. *Spokane Spokesman*, September 24, 1891; John W. Sayre, *Ghost Railroads of Central Arizona* (Phoenix: Red Rock Publishing Co., 1985), 77, 91; Philip R. Woodhouse, *Monte Cristo* (Seattle: The Mountaineers, 1996), 111–112.

20. "Wire Rope Tramways," *E&MJ* (October 3, 1903): 515; Woodhouse, *Monte Cristo*, 129–130; Allan G. Bird, *Silverton Gold: The Story of Silverton's Largest Gold Mine* (privately published, 1986), 119.

21. *Silverton Weekly Miner*, February 10, 1899; "Tramway Tower in Path of Snowslide," *E&MJ* (May 11, 1912): 933; R. D. Seymour, "Aerial Tramways in the San Juan Mountains," *The Mining and Metallurgical Journal* (November 5, 1908): 22; H. A. Morrison, "Golbe Consolidated Mill, Dedrick, Calif.," *E&MJ* (September 5, 1914): 423–424.

22. "Unusual Design of Overhead Tramway," *E&MJ* (August 16, 1913): 293; Hamilton W. Baker, "Aerial Tramway Locked by Windstorm," *E&MJ* (July 8, 1916): 91–92.

23. "Scale of wages to be paid on this property [Silver Lake Mine] on and after June 1, 1907," Duane Smith Collection; W. V. D. Camp to William C. Starr, April 11, 1918, Blue Bell Mine File (Colvocoresses), Arizona Department of Mines and Mineral Resources.

24. William C. Kuhn, "Uses and Costs of Aerial Tramways," *E&MJ* (April 7, 1917): 608–610; "Operation and Maintenance Cost of Aerial Transportation," *E&MJ* (June 3, 1916): 986–987; "General Mining News, South Dakota, Lawrence County," *E&MJ* (February 2, 1907): 254.

25. "Wire Rope Tramways," *E&MJ* (October 3, 1903): 514–515; William Hewitt, *The Bleichert System of Wire Rope Tramways* (Trenton, NJ: The Trenton Iron Company, 1908), 17; Stoiber: "Excerpts: General Report upon the Silver Lakes Mines, 1899."

26. Lundberg, *Fractured Veins and Broken Dreams*, 90; Kim K. Howell, "The History of Tiger, Arizona," typescript, Mammoth Mine File, Arizona Department of Mines and Mineral Resources.

27. John Marshall with Zeke Zanoni, *Mining the Hard Rock in the Silverton San Juans* (Silverton, CO: Simpler Way Book Co., 1996), 58; Bird, *Silverton Gold*, 124–126.

28. Rickard, *Across the San Juan Mountains*, 46: Lundberg, *Fractured Veins and Broken Dreams*, 28; *Times-Gazette*, March 5, 1898.

29. *Democrat* (Redwood City, California), August 2, 1894, quoted in David F. Myrick, "Andrew Smith Hallidie—A Remarkable Man," *La Peninsula* 17 (February 1973): 7; Metzger, "Aerial Tramways, Part 2," 39; Woodhouse, *Monte Cristo*, 64, 108.

30. Smith, *The Coeur d'Alene Mining War of 1892*, 91; Sayre, *Ghost Railroads of Central Arizona*, 91; Marshall, *Mining the Hard Rock*, 56–57.

31. Smuggler-Union tramway pass (n.d.), Idarado Mining Company Records, Ouray, Colorado; Harriet Fish Backus, *Tomboy Bride: A Woman's Personal Account of Life in Mining Camps of the West* (Boulder, CO: Pruett Publishing Co., 1977), 140–142; Russ Coleman and Dell McCoy, *The RGS Story* (Denver, CO: Sundance Books, 1991), vol. II, 377; Bird, *Silverton Gold*, 126; phone interview with Duane Smith, June 24, 1998.

32. Phone interview with Duane Smith, June 24, 1998.

CHAPTER 4

1. Rickard, *Across the San Juan Mountains*, 63. In 1908, R. D. Seymour listed the major Trenton tramways constructed in the San Juan Mountains. He noted the following lines: Hidden Treasure, Atlas, Barstow, Camp Bird, Tom Boy, Smuggler Union, Liberty Bell, Bonnie Girl, May Day, Ross, Henrietta, Gold King, Grand Mogul, Old Hundred, Green Mountain, Iowa, Arpad, Silver Lake, Gold Prince, Holy Moses, Commodore, and Amethyst. Combined, these tramways carried over five thousand tons of ore per day. In the aggregate length they extended over twenty-seven miles. See Seymour, "Aerial Tramways in the San Juan Mountains," *The Mining and Metallurgical Journal* (November 5, 1908): 22–23.

2. Duane A. Smith, *Silverton: A Quick History* (Fort Collins, CO: First Light Publishing, 1997), 48–49; Robert L. Brown, *An Empire of Silver: A History of the San Juan Silver Rush* (Caldwell, ID: The Claxton Printers, 1965), 206; William C. Prosser, "The Silver Lake Basin, Colorado," *E&MJ* (June 20, 1914): 1229–1230.

3. Prosser, "The Silver Lake Basin," 1229–1231; *Wire Rope Transportation*, 17; *Silverton Weekly Miner*, January 8, 1897.

4. Stoiber, "Excerpts: General Report upon the Silver Lake Mines, 1899"; Prosser, "The Silver Lake Basin," 1230; *Wire Rope Tramways*, 17; *Silverton Weekly Miner*, August 5, 1898.

5. Stoiber, "Excerpts: General Report upon the Silver Lake Mines, 1899"; Brown, *An Empire of Silver*, 208–209.

6. Rickard, *Across the San Juan Mountains*, 60–62.

7. Bird, *Silverton Gold*, 61–69, 118; Darlene A. Reidhead, *Tour the San Juans— Silverton to Animas Forks* (Cortez, CO: Southwest Printing Co., 1994), 217–227; *Silverton Weekly Miner*, December 31, 1897, September 30 and October 7, 1898.

8. N. C. Maxwell, "The Sunnyside Mining and Milling Company," *AX-I-DENT-AX* (September 1929): 1–10.

9. Brown, *Empire of Silver*, 216–217.

10. Ibid., 217; Joel Hoffman, "Historic Resource Evaluation of the Gold Prince Aerial Tramway," Bureau of Land Management, San Juan Resource Area, July 1990 (courtesy of Duane Smith Collection); George P. Scholl and R. L. Herrick, "The Gold Prince Mine and Mill," *Mines and Minerals* (March 1907): 340.

11. Hoffman, "Historic Resource Evaluation"; Scholl and Herrick, "The Gold Prince Mine and Mill," 340–341; H. J. Brown, "The Gold Prince Mill," *Mining Reporter* (August 30, 1906): 204–205.

12. Hoffman, "Historic Resource Evaluation."

13. Robert L. Spude, "Sound Democrat Mill," Historic American Engineering Record, CO-69, National Park Service, August 1991.

14. Sandra Dallas, *Colorado Ghost Towns and Mining Camps* (Norman: University of Oklahoma Press, 1985), 41; Rickard, *Across the San Juan Mountains*, 64; Hewitt, *Wire Rope Tramways*, 17.

15. Rickard, *Across the San Juan Mountains*, 9–12; P. David Smith, *Ouray: A Quick History* (Ouray, CO: Western Reflections, 1996), 64.

16. Ferrell, *Silver San Juan*, 118–125; Brown, *An Empire of Silver*, 303; Rickard, *Across the San Juan Mountains*, 35–38.

17. Limited information is available on many of the Colorado tramways. A few are plotted on topographical maps or cited in guide books, but very little documentary evidence remains. One exception is the Old Hundred Mine, which is featured in Seymour, "Aerial Tramways in the San Juan Mountains," 23–24.

18. William Hewitt, "Aerial Tramways at Bingham Canyon, Utah," *Mining and Engineering World* (September 7, 1912): 435.

19. Ibid.; Leroy A. Palmer, "Utah Consolidated Aerial Tramway," *Mines and Minerals* (October 1910): 151; S. S. Webber, "Wire Rope Tramway Engineering," *The Mining and Metallurgical Journal* (November 5, 1908): 5–8.

20. Hewitt, "Aerial Tramways at Bingham Canyon, Utah," 437–438.

21. Ibid., 437.

22. Ibid., 435–436; Palmer, "Utah Consolidated Aerial Tramway," *Mines and Minerals* (October 1910): 150–151; "The Highland Boy Tramway," *E&MJ* (May 28, 1910): 1108; *E&MJ* (February 19, 1910): 434.

23. "Unusual Design of Overhead Tramway," *E&MJ* (August 16, 1913): 293–296; Arthur E. Gibson, "An Aerial Tramway for Mining Cliff Coal," *Bulletin of the American Institute of Mining Engineers* (October 1914): 2537–2546.

24. Fahey, "The Brothers Riblet," *Spokane Magazine* (1980): 14; "Wire-Rope Tramway: At Grand Encampment, Wyoming, Sixteen Miles Long—Built by A. Leschen & Sons Rope Co.," *Mines and Minerals* (April 1904): 452–453.

25. "Wire-Rope Tramway: At Grand Encampment, Wyoming," 452–453; Lambert Florin, *Ghost Towns of the West* (n.p.: Promontory Press, 1970), 492–493. A good discussion of the bucket problem can be found in F. C. Carstarphen, "An Aerial Tramway for the Saline Valley Salt Company, Inyo County, California," *American Society of Civil Engineers, Transactions* 81 (December 1917): 746.

26. "Aircraft Hazards Inventory," September 5, 1997, National Park Service, Death Valley, unpublished.

27. Michael Digonnet, *Hiking Death Valley: A Guide to Its Natural Wonders and Mining Past* (Palo Alto, CA: n.p., 1997), 127–132; Philip Varney, *Southern California's Best Ghost Towns* (Norman: University of Oklahoma Press, 1990), 33–35.

28. For a detailed history of the salt tramway, see Carstarphen, "An Aerial Tramway for the Saline Valley," 709–748. See also Digonnet, *Hiking Death Valley*, 485–487; Mallory Hope Ferrell, *Southern Pacific Narrow Gauge* (Edmonds, WA: Pacific Past Mail, 1982), 74; and Mary DeDecker, *White Smith's Fabulous Salt Tram* (Death Valley, CA: The Death Valley '49ers, Inc., 1993).

29. Digonnet, *Hiking Death Valley*, 486–489; Carstarphen, "An Aerial Tramway for the Saline Valley," 728–737; Robert O. Greenawalt, "We Hiked the Inyo Bucket Brigade," *Desert Magazine* 22 (August 1959): 19–21.

30. Ferrell, *Southern Pacific Narrow Gauge*, 63, 98, 117; Varney, *Southern California's Best Ghost Towns*, 13; miscellaneous notices, *E&MJ* (May 22, 1915): 924, (July 3, 1915): 35, (September 18, 1915): 495, and (July 8, 1916): 115.

31. George M. Colvocoresses, "Report on De Soto Mine," November 28, 1945, De Soto Mines File, Arizona Department of Mines and Mineral Resources; Sayre, *Ghost Railroads of Central Arizona*, 90–91; Robert L. Spude and Stanley W. Paher, *Central Arizona Ghost Towns* (Las Vegas: Nevada Publications, 1978), 35.

32. Sayre, *Ghost Railroads of Central Arizona*, 74–75; George M. Colvocoresses, "Report on the Mines and Operations of the Consolidated Arizona Smelting Company, February 15, 1913," Annual Report, Consolidated Arizona Smelting Company, 1916; and W. V. D. Camp to W. C. Star, April 11, 1918, Blue Bell Mine File, Arizona Department of Mines and Mineral Resources.

33. "Report on De Soto Mine," Annual Report, Consolidated Arizona Smelting Company, 1916; Sayre, *Ghost Railroads of Central Arizona*, 77–78.

34. *E&MJ* (April 1, 1916): 621, (June 17, 1916): 1092, and (September 16, 1916): 515; W. Burns,"Report of the Preliminary Examination of the Gila Copper Sulphide Company, Christmas, Arizona, March 7, 1919," Christmas Mine File, Arizona Department of Mines and Mineral Resources; Myrick, *Railroads of Arizona*, vol. II, 587–592. For additional information on Arizona's tramways, see Robert A. Trennert, "Aerial Mine Tramways in Arizona," in Michael Canty, Michael Green, and H. Mason Coggin, eds., *History of Mining in Arizona*, vol. III (Tucson: Mining Foundation of the Southwest, 1999), 103–112.

35. Basque, *West Kootenay*, 45–47.

36. *Daily News* (Nelson, British Columbia), September 13, 1975; Pellowski, *Silver, Lead and Hell*, 103; N. L. Barlee, *West Kootenay: Ghost Town Country* (n.p.: Canada West Publications, 1984), 142.

37. Pellowski, *Silver, Lead and Hell*, 102–104; Dave May, *Sandon: The Mining Centre of the Silvery Slocan* (privately published, 1986), 96–98; N. L. Barlee, *Gold Creeks and Ghost Towns*, rev. ed. (Blaine, WA: Hancock House Publishers Ltd., 1984), 121.

38. May, *Sandon*, 97–98; Pellowski, *Silver, Lead and Hell*, 103.

39. John Norris, *Old Silverton, British Columbia, 1891–1930* (Silverton, B.C.: Silverton Historical Society, 1985), 143–145.

40. Lundberg, *Fractured Veins and Broken Dreams*, 28–31, 33, 37, 41, 81.

41. Ibid., 87, 90, 111–113, 120.

CHAPTER 5

1. *The Mining Catalog, Metal and Quarry Edition, 1921* (Pittsburgh: Keystone Consolidated Publishing Co., 1921), 244–245, 258–261; *The Mining Catalog, Metal and Quarry Edition, 1925* (Pittsburgh: Keystone Consolidated Publishing Co., 1925), 500–512.

2. Phone interview by Kathleen Howard with Leo Vogel, October 1, 1998; U.S. Patent #826,340 (July 17, 1906); *Pikes Peak Gold: A Photographic Account of Cripple Creek and Victor Mining District* (Victor, CO; Barbarosa Press, 1986), 25. A postcard in the author's collection dated 1912 shows the "new" Lawson-type tramway at Cripple Creek.

3. Vogel interview, October 1, 1998; Interstate Equipment Corporation advertisement, *E&MJ* (October 9, 1926).

4. *E&MJ* (April 26, 1926): 43; Blue Diamond Company brochure (ca. 1925), Interstate Equipment Corporation Collection, Pittsburgh, Pennsylvania; "Operating a Crushed Stone Plant under Unusual Conditions," *Pit and Quarry* (April 15, 1925): 41–46.

5. Gail Martin, "New Park–Utah Aerial Tramway: Capacity 1,000 Tons Daily," *E&MJ* (February 6, 1926): 254–255.

6. A. H. Hubbell, "Pecos Mine: A New Zinc-Lead Project," *E&MJ* (December 25, 1926): 1004–1012.

7. "Sidney Aerial Tramway Construction Cost Analysis, 1927," Bunker Hill Company Records (MG 367), University of Idaho Library.

8. H. S. Gieser, "Mining and Milling on Santa Catalina Island," *E&MJ* (August 13, 1927): 245–247.

9. Charles A. Chase, "A Geological Gamble in Colorado Meets with Success," *E&MJ* (August 10, 1919): 203–205; Smith, *Silverton: A Quick History*, 83–84; Robert A. Sloan and Carl A. Skowronski, *The Rainbow Route: An Illustrated History of the Silverton Railroad, the Silverton Northern Railroad, and the Silverton, Gladstone, and Northerly Railroad* (Denver: Sundance Publications Ltd., 1975), 331; Metzger, "Aerial Tramways, Part 2," 37–38; "Delay Shenandoah Mill Start," *E&MJ* (December 7, 1929): 901; "Operation of New 300-Ton Mill at Shenandoah-Dives Started," *E&MJ* (February 24, 1930): 212; Marshall, *Hard Rock*, 52–53.

10. Marshall, *Hard Rock*, 54–55, 60. Quote from page 55.

11. "Build Tramway for North Star," *E&MJ* (February 24, 1930): 212; "Empire-Star Mines Maintain Reserves as Output Rises," *E&MJ* (January 26, 2931): 79.

12. Metzger, "Aerial Tramways, Part 2," 36–39; Smith, *Silverton: A Quick History*, 83–88; Charles A. Chase, "Flotation Widely Employed in San Juan Mining," *E&MJ* (August 1935): 395–397.

13. *American Steel & Wire Company Aerial Wire Rope Tramways* (Trenton, NJ: American Steel & Wire Company, 1935), 8–9; Fred A. Carstarphen, "Improving on Rope-Tramway Locations by Using Angle Stations and Towers," *E&MJ* (January 23, 1930): 59–62; Carstarphen, "Truck or Cableway? A Comparison of Costs," *E&MJ* (July 24, 1930): 61–62; Wylie L. Graham, "World's Longest Bi-Cable Ropeway," *E&MJ* (July 1933): 287–288.

14. *American Steel & Wire Company Aerial Rope Tramways*, 84–85; "Hanging by a Wire Thread," *Mines Magazine* (September-October 1999), 5; Schneigert, *Aerial Tramways*, 11–13.

15. Smith, *Silverton: A Quick History*, 92–94.

16. Joseph Kurtak, "History of Pine Creek: A World Class Tungsten Deposit," *Mining Engineering* 50 (December 1998): 42–47; Kurtak, *Mine in the Sky: The History of California's Pine Creek Tungsten Mine and the People Who Were Part of It*, 2d ed., rev. (Anchorage, AK: Publications Consultants, 1998), 61–67; Ed Hunter to author, January 30, 2000.

17. The most complete history of the Orphan Mine is contained in a National Register of Historic Places Nomination, written in 1993 by NPS historian Harlan D. Unrau. Other useful sources include: "Field Engineer's Report," September 4, 1959, Orphan Mine File, Arziona Department of Mines and Mineral Resources; "They Go to Work in a Bucket," *Arizona Days and Ways* (June 9, 1957): 6–11; "How Western Gold Mines Uranium in Grand Canyon," *Mining World* (January 1959): 32–35; Clyde M. Brundy, "Orphan with a Midas Touch," *The Denver Post*, November 27, 1977; "Orphan Mine Suspends Uranium Ore Production," *Pay Dirt* (April 28, 1969): 26.

18. Orphan Mine National Register Nomination.

19. Ibid.
20. "Field Engineer's Report," July 29, 1957, U.S. Guano Company File, Arizona Department of Mines and Mineral Resources; *Arizona Republic*, July 28, 1957; Bud DeWald, "Canyon Cable to Riches," *Arizona Days and Ways* (January 12, 1958): 78; Greer Cheser, "Treasure of the Granite Gorge," *Cañon Journal* (spring-summer 1996): 12–13.
21. Myrick, *Railroads of Arizona*, vol. III, 288; *Phoenix Gazette*, October 6, 1972; Bill Conger, "Old Clifton Stack and Tramway Held Many Memories," *Morenci Copper Review* (May 1997): 16.
22. Sam G. Bonasso and Tom J. George, "Aerial Cableways Cut Costs on Long Steep Hauls," *Coal Mining* (November 1985): 38–41; phone interview by Kathleen Howard with Leo Vogel, October 1, 1998; James R. Barclay, "Greenex and Grouse—Two Modern Aerial Tramways," *CIM Bulletin* (July 1978): 71–77; Brian G. Pewsey et. al., "The Cassair Story," *CIM Bulletin* (April 1978): 80–87; P. Davis, "Significant Developments in Aerial Ropeway Design," *Mining Engineering* (March 1984): 241–243.
23. Lone E. Jamison, *The Copper Spike* (Anchorage: Alaska Northwest Publishing Company, 1975), 146; Robert L. Spude and Sandra M. Faulkner, eds., *Kennecott, Alaska* (Anchorage, AK: National Park Service, 1987); Robert Spude to author, January 4, 22, 1999; Steve Peterson to author, January 15, 1999.

GLOSSARY OF TRAMWAY TERMS

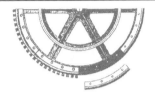

angle station   A tramway structure that accomplishes a turn in the line.

back traffic   Supplies and materials returning to a mine by tramway in the opposite direction of the ore.

bull wheel   End pulley on a tramway.

carrier (bucket)   Container attached to a tram line to hold ores or other materials.

clip   Device attached to a single-rope tramway to hold the carrier to the cable.

discharge terminal   End terminal of the tramway where the ore is unloaded.

double-rope   A tramway of the Bleichert design where the carrier rides on a track cable and is pulled by a second, smaller cable.

friction grip   A device for attaching the traction cable to the carrier on a double-rope tramway.

jig-back   Simple form of a balanced tramway that utilizes two carriers, one going up while the other goes down, and vice versa.

ropeway   Another term for an aerial tramway. Popular in the nineteenth century.

section station   Also known as a transfer station. A structure on longer tramways that transfers carriers from one section to another by means of a rail connection.

single-rope   Also known as the Endless Wire Ropeway. Definition of a tramway that uses a traveling cable with carriers attached.

tension station   Structures placed along longer tramways to apply tension to the track cables, usually at intervals of a mile or more.

traction cable   A small moving cable that pulls the carriers on a double-rope tramway.

wire rope   Steel or iron cable used on all tramways to support and move loads. Available in many sizes and surfaces.

BIBLIOGRAPHY

Manuscripts and Archival Collections

Arizona State University Library, Tempe, Arizona
  Arizona Collection
Arizona Department of Mines and Mineral Resources, Phoenix, Arizona
  Mine Records
Bancroft Library, University of California, Berkeley, California
  William Miles Read Papers
California State Archives, Sacramento, California
  Secretary of State Records
Cheney Cowles Museum, Spokane, Washington
  Riblet Company Files
Nevada Historical Society, Reno, Nevada
  Miscellaneous Records
New Jersey State Archives, Trenton, New Jersey
  Department of State Records
  Miscellaneous Records
New York Historical Society, New York City, New York
  Olcott Papers

Northern Mine Research Society, Keighley, Yorkshire, England
  Miscellaneous Records
Private Collections
  Duane A. Smith Collection, Durango, Colorado
  Ed Hunter Collection, Victor, Colorado
  Idarado Mining Company Collection, Ouray, Colorado
  Interstate Equipment Corporation Collection, Pittsburgh, Pennsylvania
  Robert A. Trennert Collection, Chandler, Arizona
  Robert L. Spude Collection, Santa Fe, New Mexico
State of Missouri, Jefferson City
  Secretary of State Records
University of Idaho Library, Moscow, Idaho
  Bunker Hill Company Records (MG 367)
  Hercules Mining Company Records (MG 236)

GOVERNMENT DOCUMENTS

FEDERAL

Hoffman, Joel. *Historic Resource Evaluation of the Gold Prince Aerial Tramway*, Bureau of Land Management, San Juan Resource Area, July 1990.

Metzger, O. H. *Aerial Tramways in the Metal-Mining Industry, Part 1*. U.S. Bureau of Mines, Information Circular #6948, September 1937.

———. *Aerial Tramways in the Metal-Mining Industry, Part 2, Construction and Operating Costs*. U.S. Bureau of Mines, Information Circular #7095, February 1940.

Skeirik, Ronald, and C. Taylor. *Chilkoot Pass Route Tramways*. Historic American Engineering Record AK-13 (blueprints), National Park Service, 1986–1987.

Spude, Robert L. *Sound Democrat Mill*. Historic American Engineering Record CO-69, National Park Service, 1991.

Unrau, Harlan D. *Orphan Mine*. National Register of Historic Places Nomination, National Park Service, 1993.

U.S. Patent Office. *Official Gazette of the United States Patent Office*, vol. 107. Washington, DC: Government Printing Office, 1904.

———. *United States Patents*. Research Publications, Inc. (microfilm).

STATE

Acts of the 71st Legislature of the State of New Jersey. Trenton, NJ: Phillips and Boswell, 1847.

People ex. rel. Aspen M. & S. Co. v. District Court of Pitkin County. Colorado Reports 11 (1887), 147–156.

Report of the Mineralogist of the State of Nevada for the Years 1869 and 1870. Carson City, NV: Charles L. Perkins, 1871.

MANUFACTURERS' CATALOGS

American Steel & Wire Company Aerial Wire Rope Tramways. Trenton, NJ: American Steel & Wire Company, 1935.

Ballard, C. M. *Cutting Material Handling Costs*. Williamsport, PA: Williamsport Wire Rope Company, 1929.

[Hallidie, Andrew S.]. *Transportation of Ore and Other Material by Means of Endless Traveling Wire Ropes*. San Francisco: C. A. Murdock & Company, 1878.

Hewitt, William. *The Bleichert System of Aerial Tramways*. Trenton, NJ: Trenton Iron Company, 1910.

———. *The Bleichert System of Wire Rope Tramways*. Trenton, NJ: Trenton Iron Company, 1908.

Huson's Patent Automatic Wire Rope Tramway. Denver: C. W. Badgely & Company, [1893].

Otto Aerial Tramways. Chicago: Fraser & Chalmers, 1898.

Riblet Aerial Tramways. Spokane, WA: Shaw and Borden, [1930].

Roebling Wire Rope and Wire, Pacific Coast Edition. Trenton, NJ: John A. Roebling's Sons, 1931.

The Mining Catalog, Metal and Quarry Edition, 1921. Pittsburgh: Keystone Consolidated Publishing Company, 1921.

The Mining Catalog, Metal and Quarry Edition, 1925. Pittsburgh: Keystone Consolidated Publishing Company, 1925.

Wire Rope and Its Applications. St. Louis: Broderick & Bascom Company, 1939.

Wire Rope Tramways for Economical Transportation: Bleichert Patent System. New York: Cooper, Hewitt & Company, 1888.

Wire Rope Tramways with Special Reference to the Bleichert Patent System. Trenton, NJ: Trenton Iron Company, 1890.

Wire Rope Transportation in All Its Branches. Trenton, NJ: Trenton Iron Company, 1896.

MINING AND PROFESSIONAL JOURNALS

Numerous articles and notices appeared in a variety of mining and professional journals. These articles are identified specifically in the notes. Listed below are the journals consulted:

American Institute of Mining Engineers, Transactions
American Society of Civil Engineers, Transactions
AX-I-DENT-AX
Bulletin of the American Institute of Mining Engineers
Cassier's Magazine
CIM Bulletin
Coal Mining
Engineering and Mining Journal
Engineering News
Iron Age
Mines Magazine
Mines and Minerals
Mining and Engineering World

Mining and Metallurgical Journal
Mining and Scientific Press
Mining Engineering
Mining Reporter
Mining World
Pay Dirt
Pit and Quarry
Railroad Gazette
Scientific American

NEWSPAPERS

Arizona Silver Belt (Globe, Arizona)
Daily News (Nelson, British Columbia)
Democrat (Redwood City, California)
Denver Post (Denver, Colorado)
Phoenix Gazette (Phoenix, Arizona)
Silverton Weekly Miner (Silverton, Colorado)
Spokane Spokesman (Spokane, Washington)
Spokane Review (Spokane, Washington)
Spokesman Review (Spokane, Washington)
Territorial Enterprise (Virginia City, Nevada)
Times-Gazette (Redwood City, California)

BOOKS

Angel, Myron, ed. *History of Nevada*. Oakland: Thompson and West, 1881.

Backus, Harriet Fish. *Tomboy Bride: A Woman's Personal Account of Life in Mining Camps of the West*. Boulder, CO: Pruett Publishing Company, 1977.

Bailey, Lynn R. *Supplying the Mining World: The Mining Equipment Manufacturers of San Francisco, 1850–1900*. Tucson, AZ: Westernlore Press, 1996.

Barlee, N. L. *Gold Creeks and Ghost Towns*. Rev. ed. Blaine, WA: Hancock House Publishers, Ltd., 1984.

———. *West Kootenay: Ghost Town Country*. N.p.: Canada West Publications, 1984.

Basque, Garnet. *West Kootenay: The Pioneer Years*. Langley, B.C.: Sunfire Publications, 1990.

Bird, Allan G. *Silverton Gold: The Story of Colorado's Largest Gold Mine*. Privately published, 1986.

Brown, Robert L. *An Empire of Silver: A History of the San Juan Silver Rush*. Caldwell, ID: The Caxton Printers, 1965.

Canty, Michael, Michael Green, and H. Mason Coggin, eds. *History of Mining in Arizona*, vol. III. Tucson, AZ: Mining Foundation of the Southwest, 1999.

Coleman, Russ, and Dell A. McCoy. *The RGS Story*, vol. II. Denver: Sundance Books, 1991.

Coleman, Russ, Dell A. McCoy, and William A. Graves. *The RGS Story*, vols. III and IV. Denver: Sundance Books, 1993, 1994.

Dallas, Sandra. *Colorado Ghost Towns and Mining Camps*. Norman: University of Oklahoma Press, 1985.

DeDecker, Mary. *White Smith's Fabulous Salt Tram*. Death Valley, CA: The Death Valley '49ers, Inc., 1993.

Digonnet, Michael. *Hiking Death Valley: A Guide to Its Natural Wonders and Mining Past*. Palo Alto, CA: n.p., 1997.

Ferrell, Mallory Hope. *Silver San Juan: The Rio Grande Southern*. Boulder, CO: Pruett Publishing Company, 1973.

———. *Southern Pacific Narrow Gauge*. Edmonds, WA: Pacific Fast Mail, 1982.

Florin, Lambert. *Ghost Towns of the West*. N.p.: Promontory Press, 1970.

Jackson, W. Turrentine. *Treasure Hill: Portrait of a Mining Camp*. Tucson: University of Arizona Press, 1963.

Jamison, Lone E. *The Copper Spike*. Anchorage: Alaska Northwest Publishing Company, 1975.

Kahn, Edgar M. *Andrew Smith Hallidie: A Tribute to a Pioneer California Industrialist*. San Francisco: privately published, 1953.

Kent William. *Kent's Mechanical Engineer's Handbook*. New York: John Wiley & Sons, 1923.

Kurtak, Joseph M. *Mine in the Sky: The History of California's Pine Creek Tungsten Mine and the People Who Were Part of It*. 2d ed, rev. Anchorage, AK: Publication Consultants, 1998.

Lindley, Curtis H. *A Treatise on the Law Relating to Mines and Mineral Lands*. 3d ed. San Francisco: Bancroft-Whitney Company, 1914.

Lundberg, Murray. *Fractured Veins and Broken Dreams: Montana Mountain and the Windy Arm Stampede*. Whitehorse, Yukon Territory: Pathfinder Publications, 1996.

Mack, Edward C. *Peter Cooper, Citizen of New York*. New York: Duell, Sloan and Pearce, 1949.

Marshall, John, with Zeke Zanoni. *Mining the Hard Rock in the Silverton San Juans: A Sense of Place, a Sense of Time*. Silverton, CO: Simpler Way Book Company, 1996.

May, Dave. *Sandon: The Mining Centre of the Silvery Slocan*. Privately published, 1986.

Myrick, David F. *Railroads of Arizona*. Vol. III, *Clifton, Morenci and Metcalf, Rails and Copper Mines*. Glendale, CA: Howell-North Books, 1980.

———. *Railroads of Nevada and Eastern California*. Vol. I, *The Northern Roads*. Berkeley, CA: Howell-North Books, 1962.

Norris, John. *Old Silverton, British Columbia, 1891–1930*. Silverton, B.C.: Silverton Historical Society, 1985.

Nossaman, Allen. *Many More Mountains*. Vol. III, *Rails into Silverton*. Denver: Sundance Books, 1998.

Paul, Rodman W. *Mining Frontiers of the Far West, 1848–1880.* New York: Holt, Rinehart and Winston, 1963.

Peele, Robert, ed. *Mining Engineer's Handbook.* 2 vols. New York: John Wiley & Sons, 1927.

Pellowski, Veronika. *Silver, Lead and Hell: The Story of Sandon.* Sandon, B.C.: Prospector's Pick Publishing, 1992.

Pikes Peak Gold: A Photographic Account of Cripple Creek and Victor Mining Dictrict. Victor, CO: Barbarosa Press, 1986.

Raymond, Rossiter W. *Peter Cooper.* Freeport, NY: Books for Libraries Press, 1972. Reprint of 1901 edition.

Reidhead, Darlene A. *Tour the San Juans—Silverton to Animas Forks.* Cortez, CO: Southwest Printing Company, 1994.

Rickard. T. A. *Across the San Juan Mountains.* New York: Engineering and Mining Journal, 1903.

Sayre, John W. *Ghost Railroads of Central Arizona.* Phoenix: Red Rock Publishing Company, 1985.

Schneigert, Zbignieu. *Aerial Tramways and Funicular Railways.* Warsaw, Poland: Pergamon Press, 1966.

Sloan, Robert A., and Carl A. Skowronski. *The Rainbow Route: An Illustrated History of the Silverton Railroad, the Silverton Northern Railroad, and the Silverton, Gladstone, and Northerly Railroad.* Denver: Sundance Publications Ltd., 1975.

Smith, David. *Ouray: A Quick History.* Ouray, CO: Western Reflections, 1996.

Smith, Duane A. *Silverton: A Quick History.* Fort Collins, CO: First Light Publishing, 1997.

Smith, Robert Wayne. *The Coeur d'Alene Mining War of 1892: A Case Study of an Industrial Dispute.* Corvallis: Oregon State University Press, 1961.

Spence, Clark C. *British Investments and the American Mining Frontier, 1860–1901.* Ithaca, NY: Cornell University Press, 1958.

Spude, Robert L., ed. *Chilkoot Trail: Historical Data.* Fairbanks: University of Alaska Press, 1980.

Spude, Robert L., and Sandra M. Faulkner, eds. *Kennecott, Alaska.* Anchorage, AK: National Park Service, 1987.

Spude, Robert L., and Stanley W. Paher. *Central Arizona Ghost Towns.* Las Vegas: Nevada Publications, 1978.

Steinman, D. B. *The Builders of the Bridge: The Story of John Roebling and His Son.* New York: Harcourt, Brace and Company, 1945.

Varney, Philip. *Southern California's Best Ghost Towns.* Norman: University of Oklahoma Press, 1990.

Wallace-Taylor, Alexander J. *Aerial or Wire Rope-Ways: Their Construction and Management.* London: Crosby Lockwood and Sons, 1911.

Woodhouse, Philip R. *Monte Cristo.* Seattle: The Mountaineers, 1996.

Young, Otis E., Jr. *Western Mining.* Norman: University of Oklahoma Press, 1970.

ARTICLES

Barr, William A. "Man against the Corporations." *Pacific Northwesterner* 31 (1987): 57–64.

Cheser, Greer. "Treasure of Granite Gorge." *Cañon Journal* 2 (spring-summer 1996): 4–13.

Conger, Bill. "Old Clifton Stack and Tramway Held Many Memories." *Morenci Copper Review* (May 1997): 16.

DeWald, Bud. "Canyon Cable to Riches." *Arizona Days and Ways* (January 12, 1958): 78.

Fahey, John. "The Brothers Riblet." *Spokane Magazine* 4 (November 1980): 14–17.

Greenawalt, Robert O. "We Hiked the Inyo Bucket Brigade." *Desert Magazine* 22 (August 1959): 19–21.

Jackson, W. Turrentine. "The Infamous Emma Mine: A British Interest in the Little Cottonwood District, Utah Territory." *Utah Historical Quarterly* 23 (October 1955): 339–362.

Kahn, Edgar M. "Andrew Smith Hallidie." *California Historical Society Quarterly* 19 (June 1940): 144–156.

Myrick, David A. "Andrew Smith Hallidie—A Remarkable Man." *La Peninsula* 17 (February 1973): 1–8.

Smith, Duane A. "'Where a Bird Could Hardly Obtain a Footing': George Armstrong Custer and the Stevens Mine." *Colorado Heritage* 17 (spring 1997): 25–35.

"They Go to Work in a Bucket." *Arizona Days and Ways* (June 9, 1957): 6–11.

Trennert, Robert A. "From Gold Ore to Bat Guano: Aerial Mine Tramways in the West." *The Mining History Journal* 4 (1997): 3–13.

UNPUBLISHED MATERIAL

"Aircraft Hazards Inventory." Death Valley, CA: National Park Service, 1997.

Norris, Frank. "The Tramway Story." Karl Gurcke, ed. Skagway, AK: Klondike Gold Rush National Historical Park, n.d.

CORRESPONDENCE AND INTERVIEWS

Hunter, Ed, to author, January 30, 2000.

Peterson, Steve, to author, January 15, 1999.

Spude, Robert L., to author, January 4, 22, 1999.

Smith, Duane, phone interview with author, June 24, 1998.

Vogel, Leo, phone interview with Kathleen Howard, October 1, 1998.

INDEX

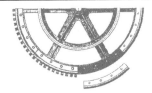

Page numbers in italics indicate illustrations.

Grassman, Edward J.: tramways by, 93

Grass Valley gold district, 98

Gravity power, 12, 56

Gray, Joshua, wire rope factory of, 20

Green Mountain Mine, tramway at, 74

Guadalajara, tramway near, 26

Guano, 102

Hallidie, Andrew Smith, *10*, 22; cable car system and, 14; challengers for, 20–21; components by, *13*; design problems for, 30; improved grip pulley and, 11; inventions by, 12–13, 19; tramways by, 9, 11, 17–21, *21*, 27, 29, 44, 61, 85

Hallidie's Endless Wire Ropeway, 11; advertisements for, 13–14; booklet on, 20

Hamilton, described, 14

Hammond, John Hays, 36, 37

Hanson, Rasmus, Huson tramway and, 24–25

Hard-rock mining, 2, 8

Harley Mine, tramway at, 20

Harris, George, wire rope and, 3

Hayden's Peak, snowslide at, 58

Henrietta Mine, tramway at, 74

Hercules Mining Company, legal advice for, 49

Hewitt, Abram S., Trenton Company and, 4

Hewitt, William: on tramways, 42, 43, 75

Hewitt–Lorna Doone Mine, lines at, 87

Hewitt Sampler, 49

Highland Boy Mine, 74; tramway at, 75, 99

Hodgson, Charles, 11, 30; tramways by, 14, *15*

Hogan, Daniel L. "Pops," 101

Holy Moses Mine, bucket line at, 38

Human cargo, transporting, 60–62, *61*, *62*, 63

Huson, Charles M.: tramways by, 21, 24, 27, 29

Huson tramways, *23*, 26, 38, 39, 42, 44, 69; components of, *24*; described, 22; installing, 24–25

Hydraulic mining, 8

Injured miners, transporting, 63

Inspiration Copper Company, 85

Intermural Railway, 38

International Mill, tramway at, 15, 16

International Smelting and Refining Company, 75

Interstate Equipment Corporation, 92, 105; Continental Tramway and, 93; tramway by, 103–104

Iowa Gold Mining and Milling Company, 67

J. H. Montgomery Machinery Company, technology by, 39

Jig-back tramway, 74, 102, *103*

John A. Roebling's Sons Company, 21; cables from, 95; tramway by, 40

J. Pohlig Company, 38, 40

Keane Wonder Gold Mining Company, tramway of, 79–80, 108

Keeler, 81, 82, *83*

Kelsey, Martha A., 43, 60–61

Kennecott Mines, 107

Kinney, W. Z., construction by, 71

Klondike gold rush, 41

Lanyon, R. E., 88, 89

Lawrence, Frank, tramways by, 93

Lawson, William C., tramway by, 92–93

Lawson system, *92*, 93, *94*, 103–104

Leistershire stone quarries, cableway at, 11

Le Roi Mine and Museum, tramway at, 85, 105

Liberal Bell Mine, tramway at, 66

Lintorf Mines and Smelting Works, 30

Little Cottonwood Canyon; fraud in, 19–20; wire ropeway for, 18–19

Little Giant Mine, mill at, 66

Little Giant Peak, 69

Little Nation Mill, ruins of, *107*

Little Nation Mine, 69

Locked Coil Cables, 53

Lode mining, 8

Loop-Line Tramway, described, 93